Engineering Systems Reliability, Safety, and Maintenance

An Integrated Approach

Engineering Systems Reliability, Safety, and Maintenance
An Integrated Approach

B.S. Dhillon

CRC Press
Taylor & Francis Group
Boca Raton London New York

CRC Press is an imprint of the
Taylor & Francis Group, an **informa** business

CRC Press
Taylor & Francis Group
6000 Broken Sound Parkway NW, Suite 300
Boca Raton, FL 33487-2742

First issued in paperback 2019

ISBN-13: 978-0-4987-8163-3 (hbk)
ISBN-13: 978-0-367-88998-2 (pbk)

Library of Congress Cataloging-in-Publication Data

Names: Dhillon, B.S. (Balbir S.), 1947- author.
Title: Engineering systems reliability, safety, and maintenance : an integrated
approach / B.S. Dhillon.
Description: Boca Raton : Taylor & Francis, a CRC title, part of the Taylor & Francis
imprint, a member of the Taylor & Francis Group, the academic division of T&F
Informa, plc, [2017] | Includes bibliographical references and index.
Identifiers: LCCN 2016040251| ISBN 9781498781633 (hardback : alk. paper) |
ISBN 9781498781640 (ebook)
Subjects: LCSH: Engineering systems. | Reliability (Engineering) | System safety. |
Engineering systems--Maintenance and repair.
Classification: LCC TA168 .D52 2017 | DDC 620.001/1--dc23
LC record available at https://lccn.loc.gov/2016040251

Visit the Taylor & Francis Web site at
http://www.taylorandfrancis.com

and the CRC Press Web site at
http://www.crcpress.com

This book is affectionately dedicated to the memory

of late family friend Narinder Singh.

Contents

Preface

Today, engineering systems are an important element of the world economy, and each year, billions of dollars are spent to develop, manufacture, operate, and maintain various types of engineering systems around the globe. Many of these systems are highly sophisticated and contain millions of parts. For example, a Boeing jumbo 747 is made up of approximately 4.5 million parts including fasteners. Needless to say, reliability, safety, and maintenance of systems such as this have become more important than ever before. Global competition and other factors are forcing manufacturers to produce highly reliable, safe, and maintainable engineering products.

It means that there is a definite need for reliability, safety, and maintenance professionals to work closely during design and other phases. To achieve this goal, it is essential that they have an understanding of each other's discipline to a certain degree. At present, to the best of the author's knowledge, there is no book that covers the topics of reliability, safety, and maintenance within its framework. It means, at present, to gain knowledge of each other's specialties, these specialists must study various books, reports, or articles on each of the topics in question. This approach is time consuming and rather difficult because of the specialized nature of the material involved.

Thus, the main objective of this book is to combine these three topics into a single volume and to eliminate the need to consult many diverse sources in obtaining basic and up-to-date desired information on the topics. The sources of most of the material presented are given in the reference section at the end of each chapter. This will be useful to readers if they desire to delve more deeply into a specific topic or area. The book contains a chapter on mathematical concepts and another chapter on the basics of reliability, safety, and maintenance considered useful to understand the contents of subsequent chapters. Furthermore, another chapter is devoted to methods considered useful to analyze the reliability, safety, and maintenance of engineering systems.

The topics covered in the book are treated in such a manner that the reader will require no previous knowledge to understand the contents. At appropriate places, the book contains examples along with their solution, and there are numerous problems at the end of each chapter to test the reader's comprehension in the area. An extensive list of publications dating from 1926 to 2013, directly or indirectly on engineering systems reliability, safety, and maintenance, is provided at the end of this book to give readers a view of the intensity of developments in the area.

The book is composed of 11 chapters. Chapter 1 presents the need for and the historical developments in reliability, safety, and maintenance; engineering systems reliability/safety/maintenance-related facts, figures, and examples; important terms and definitions; and useful sources for obtaining information on reliability, safety, and maintenance. Chapter 2 reviews mathematical concepts considered useful to understand subsequent chapters. Some of the topics covered in the chapter are Boolean algebra laws, probability properties, statistical distributions, and useful mathematical definitions.

Chapter 3 presents various introductory aspects of reliability, safety, and maintenance. Chapter 4 presents a number of methods considered useful to analyze engineering systems reliability, safety, and maintenance. These methods are fault tree analysis, the Markov method, failure modes and effect analysis, probability tree analysis, technique of operations review, hazard and operability analysis, interface safety analysis, maintenance program effectiveness evaluation approach for managers, and indices for maintenance management analysis. Chapter 5 is devoted to computer, Internet, and robot system reliability. Some of the topics covered in the chapter are computer failure sources, computer-related faults classifications and reliability measures, fault masking, Internet failure examples, a method for automating fault detection in Internet services, categories of robot failures, and robot reliability measures and analysis methods.

Chapter 6 is devoted to transportation system failures and human errors in transportation systems and covers topics such as defects in vehicle parts and categories of vehicle failures, rail weld failures and defects, rail tanker failure modes and causes of failures, mechanical failure-related aviation accidents, ship failures, typical human error occurrence areas in railway operation, types of pilot–controller communication-related errors, methods for reducing the manning impact on shipping system reliability, and common driver errors. Chapter 7 presents various important aspects of software, robot, and transportation system safety. Some of the topics covered in the chapter are software safety assurance program; software hazard analysis methods; robot safety-related problems causing weak points in planning, design, and operation; robot safeguard methods; truck and bus safety-related issues; railroad tank safety; analysis of world airline accidents; and marine accidents.

Chapter 8 is devoted to medical and mining systems safety. Some of the topics covered in the chapter are medical system safety-related facts and figures, types of medical device/system safety, safety in medical device/system life cycle, methods for conducting medical device/system safety analysis, mining equipment/systems safety-related facts and figures, causes for mining equipment-related accidents, mining equipment maintenance-related accidents, and methods for performing mining equipment/system safety analysis. Chapter 9 is devoted to software maintenance and reliability-centered maintenance and covers topics such as software maintenance problems and maintenance types, software maintenance methods, software maintenance costing, reliability centered maintenance goals and principles, reliability-centered maintenance process, elements of reliability-centered maintenance, and reliability-centered maintenance program effectiveness measurement indicators.

Chapter 10 presents various important aspects of maintenance safety and human error in aviation and power plant maintenance. Some of the topics covered in the chapter are maintenance safety-related facts, figures, and examples; factors responsible for dubious safety reputation in performing maintenance tasks and reasons for safety-related problems in maintenance; maintenance personnel safety; guidelines for equipment/system designers for improving safety in maintenance; causes of human error in aviation maintenance; common human errors in aircraft maintenance tasks; methods for performing aircraft maintenance error analysis; human error causes in power plant maintenance and most susceptible maintenance tasks to human error in power generation; and guidelines to reduce and prevent human error in power generation maintenance. Finally, Chapter 11 presents six mathematical models for performing engineering system reliability, safety, and maintenance analysis.

This book will be useful to many individuals, including design engineers; system engineers, reliability and safety professionals; maintenance engineers; engineering administrators; graduate and senior undergraduate students in the area of engineering; researchers and instructors of reliability, safety, and maintenance; and engineers-at-large.

I am deeply indebted to many individuals, including family members, colleagues, friends, and students for their invisible inputs. The invisible contributions of my children are also appreciated. Last, but not least, I thank my wife Rosy, my other half and friend, for typing this entire book and for her timely help in proofreading.

B.S. Dhillon
Ottawa, Ontario

Author

B.S. Dhillon, PhD, is a professor of engineering management in the Department of Mechanical Engineering at the University of Ottawa, Ottawa, Canada. He has served as a chairman/director of the Mechanical Engineering Department/Engineering Management Program for more than 10 years at the same institution. He is the founder of the probability distribution named *Dhillon distribution/law/model* by statistical researchers in their publications around the world. He has published more than 70 single-authored and 160 coauthored journal articles and 152 submissions to conference proceedings on reliability engineering, maintainability, safety, engineering management, and other topics. He is or has been on the editorial boards of 12 international scientific journals. In addition, Dr. Dhillon has written 44 books on various aspects of healthcare, engineering management, design, reliability, safety, and quality published by Wiley (1981), Van Nostrand (1982), Butterworth (1983), Marcel Dekker (1984), and Pergamon (1986). His books are used in over 100 countries, and many of them are translated into languages such as German, Russian, Chinese, and Persian (Iranian).

He has served as the general chairman of two international conferences on reliability and quality control held in Los Angeles and Paris in 1987. Dr. Dhillon has also served as a consultant to various organizations and bodies and has many years of experience in the industrial sector. He has lectured in over 50 countries, including keynote addresses at various international scientific conferences held in North America, Europe, Asia, and Africa. In March 2004, Dr. Dhillon was a distinguished speaker at the Conference/Workshop on Surgical Errors (sponsored by the White House Health and Safety Committee and the Pentagon), held at the Capitol (Washington, DC).

Dr. Dhillon attended the University of Wales, where he earned a BS in electrical and electronic engineering and an MS in mechanical engineering. He earned his PhD in industrial engineering from the University of Windsor.

1

Introduction

1.1 Background

The history of the reliability field may be traced back to the early 1930s when probability concepts were applied to electric power generation-related problems [1,2]. During World War II, Germans applied the basic reliability concepts for improving reliability of their V1 and V2 rockets. During the period of 1945–1950, the US Department of Defense performed various studies concerning electronic equipment failure, equipment maintenance, repair cost, etc. As the result of these studies, it formed an ad hoc committee on reliability, and in 1952, the committee was transformed to a permanent body: the Advisory Group on the Reliability of Electronic Equipment. A detailed history of the reliability field is available in the study of Dhillon [3].

The history of the safety field may be traced back to the Code of Hammurabi (2000 BC) developed by a Babylonian ruler named Hammurabi. In modern times, in 1868, a patent was awarded for the first barrier safeguard in the United States [4]. In 1893, the US Congress passed the Railway Safety Act, and in 1912, the Cooperative Safety Congress met in Milwaukee, Wisconsin [4,5]. Additional information on the history of safety is available in the study of Dhillon [6].

Although humans have felt the need for maintaining their equipment since the beginning of time, the beginning of modern engineering equipment/system maintenance may be regarded as the development of the steam engine, in 1769, in the United Kingdom, by James Watt (1736–1819) [7]. In the United States, a magazine entitled *Factory*, which first appeared in 1882, has played an important role in the development of the engineering systems/equipment maintenance field [8]. A book on maintenance of railways was published in 1886 [9].

Needless to say, each year, billions of dollars are spent on engineering system reliability, safety, and maintenance around the world and engineering system reliability, safety, and maintenance has become a very important issue.

Over the years, a large number of publications directly or indirectly related to engineering system reliability, safety, and maintenance have appeared in the form of journal articles, conference proceeding articles, technical reports, etc. A list of over 335 such publications is provided in the Appendix section.

1.2 Engineering System Reliability, Safety, and Maintenance Facts, Figures, and Examples

Some of the facts, the figures, and the examples, directly or indirectly, concerned with engineering system reliability/safety/maintenance are as follows:

- Each year, the US industry spends around US$300 billion on plant maintenance and repair [10].
- As per Kane [11], in 1996, the direct cost of corrosion-related failures including maintenance in the US petroleum industry was US$3.7 billion per year.
- As per Backtrom [12], some studies performed in Japan clearly indicate that more than 50% of working accidents with robots can be attributed to faults in the control system's electronic circuits.
- As per Herrmann [13] and Kletz [14], the number of persons killed because of computer system-related failures was somewhere between 1000 and 3000.
- In 1974, Turkish Airlines Flight 981 (aircraft type: McDonnell Douglas DC-10-10) crashed because of cargo hatch failure and control cable failures and caused 346 fatalities [15].
- In 2002, an Amtrak auto train derailed because of malfunctioning brakes and poor track maintenance near Crescent City, Florida, and caused four fatalities and 142 injuries [16].
- In 1991, United Airlines Flight 585 (aircraft type: Boeing 737-291) crashed because of rudder device malfunction and caused 25 fatalities [17].
- As per Dhillon [18], the Emergency Care Research Institute after examining a sample of 15,000 hospital products concluded that 4–6% of these products were dangerous enough for warranting immediate corrective action.
- In 2002, a study commissioned by the National Institute of Standards and Technology reported that software errors cost the US economy approximately US$59 billion per year [19].
- The Internet has grown from 4 hosts in 1969 to over 147 million hosts and 38 sites in 2002, and in 2001, there were over 52,000 Internet-associated failures and incidents [20,21].

- As per Charette [22], over 80% of a software product's life is spent in maintenance.
- As per Fairley [23], a software product's typical life span is 1–3 years in development and 5–15 years in use (maintenance).
- For the period of 1978–1987, there were 10 robot-related fatal accidents in Japan [24].
- A study reported that 12–17% of the accidents in the industrial sector using advanced manufacturing technology were related to automated production equipment [25,26].
- Maintenance error contributes approximately 15% of air carrier accidents, and annually, it costs the US industrial sector over US$1 billion [27].
- A study of safety-associated issues concerning onboard fatalities of jet fleets worldwide for the period of 1982–1991 revealed that inspection and maintenance were clearly the second most important safety issue, with a total of 1481 onboard fatalities [28,29].
- A study by the US Nuclear Regulatory Commission reported that about 65% of nuclear system failures involve human error [30].
- Maintenance errors account for approximately 60% of the annual power loss due to human errors in fossil power plants [31].
- As per Varma [32], during the period of 1990–1994, approximately 27% of the commercial nuclear power plant outages in the United States were the result of human error.
- A study of over 4400 maintenance-associated records concerning a boiling water reactor nuclear power plant covering the period from 1992 to 1994 revealed that approximately 7.5% of all failure records could be attributed to human errors related to maintenance activities/tasks [33,34].
- The US government spends around 40% of the total software-related cost on maintenance [35].
- A Boeing study revealed that about 19.2% of in-flight engine shutdowns are due to maintenance error [27].
- A study reported that around 18% of all aircraft accidents are maintenance related [36,37].
- In 1979, in a DC-10 aircraft accident in Chicago, 272 persons lost their lives because of incorrect procedures followed by maintenance personnel [38].
- In coal mining operations throughout the United States, during the period of 1990–1999, 197 equipment fires caused 76 injuries [39].
- As per Burgess-Limerick and Steiner [40], in 2004, approximately 17% of the 37,445 injuries in US coal mines were directly or indirectly associated with bolting machines.

1.3 Terms and Definitions

There are a large number of terms and definitions used in the area of engineering system reliability, safety, and maintenance. Some of these are presented in the following [41–45]:

- *Reliability:* The probability that an item will perform its stated mission satisfactorily for the specified period when used according to the stated conditions.
- *Safety:* The conservation of human life and the prevention of damage to items as per mission requirements.
- *Maintenance:* All actions appropriate to retain an item/equipment in, or restoring it to, a given condition.
- *Availability:* The probability that an item/system is available for application or use when required.
- *Downtime:* The time period during which the item/system is not in a condition to carry out its stated mission.
- *Mission time:* The element of uptime that is needed to perform a specified mission profile.
- *Failure:* The inability of an item to function within the specified guidelines.
- *Continuous task:* A task that involves some kind of tracking activity (e.g., monitoring a changing situation/condition).
- *Safety management:* The accomplishment of safety through the effort of other personnel.
- *Unsafe condition:* Any condition, under the right set of conditions that will result in an accident.
- *Hazard:* The source of energy and the physiological and behavioral factors which, when uncontrolled properly, lead to harmful occurrences.
- *Safety process:* A series of procedures followed for enabling all safety-associated requirements of an item to be identified and satisfied.
- *Corrective maintenance:* The unscheduled repair/maintenance for returning items/equipment to a defined state and carried out because maintenance persons or users perceived failures/ deficiencies.
- *Safeguard:* A barrier guard, a device, or a procedure developed to protect humans.
- *Redundancy:* The existence of more than one means for carrying out a specified function.

- *Preventive maintenance:* All actions carried out on a planned, periodic, and specific schedule to keep an item/equipment in the stated operating condition through the process of checking and reconditioning. These actions are precautionary steps undertaken for forestalling or lowering the probability of failure or an unacceptable level of degradation in later service, rather than correcting them after their occurrence.
- *Human error:* The failure to perform a specified task (or the performance of a forbidden action) that could result in the disruption of scheduled operations or result in damage to property and equipment.
- *Predictive maintenance:* The use of modern measurement and signal-processing approaches for accurately diagnosing equipment/item condition during operation.
- *Inspection:* The qualitative observation of an item's condition/ performance.
- *Overhaul:* Comprehensive inspection and restoration of an item/ equipment to an acceptable level at a durability time/usage limit.
- *Failure mode:* The abnormality of system/item performance that causes the system/item to be considered as failed.
- *Human reliability:* The probability of accomplishing a task successfully by humans at any required stage in system operation. In some situations, the task must be accomplished within a specified time limit.

1.4 Useful Sources for Obtaining Information on Reliability, Safety, and Maintenance

There are many sources for obtaining information, directly or indirectly, concerned with engineering system reliability, safety, and maintenance. Some of the sources considered most useful are presented in the following sections under a number of distinct categories.

1.4.1 Organizations

- Reliability Society, Institute of Electrical and Electronics Engineers (IEEE), PO Box 1331, Piscataway, New Jersey
- System Safety Society, 1452 Culver Drive, Suite A-261, Irvine, California
- American Society of Safety Engineers, 1800 East Oakton Street, Des Plaines, Illinois

- National Safety Council, 444 North Michigan Avenue, Chicago, Illinois
- US Consumer Product Safety Commission, Washington, DC
- British Safety Council, 62 Chancellors Road, London, United Kingdom
- Occupational Safety and Health Administration, US Department of Labor, 200 Constitution Avenue, Washington, DC
- Civil Aviation Safety Authority, North Bourne Avenue and Barry Drive Intersection, Canberra, Australia
- Society for Maintenance and Reliability Professionals, 401 North Michigan Avenue, Chicago, Illinois
- Maintenance Engineering Society of Australia (a Technical Society of the Institution of Engineers, Australia), 11 National Circuit, Barton, Australia
- Japan Institute of Plant Maintenance, Shuwa Shiba-Koen 3-Chome Building, 3-1-38 Shiba-Koen, Minato-Ku, Tokyo, Japan

1.4.2 Journals and Magazines

- *Reliability Engineering and System Safety*
- *IEEE Transactions on Reliability*
- *Microelectronics and Reliability*
- *International Journal of Reliability, Quality, and Safety Engineering*
- *National Safety News*
- *Journal of Quality in Maintenance Engineering*
- *Maintenance Journal*
- *Journal of Safety Research*
- *Accident Analysis and Prevention*
- *Industrial Maintenance and Plant Operation*
- *Maintenance and Asset Management Journal*
- *Safety Management Journal*
- *Reliability: The Magazine for Improved Plant Reliability*
- *Maintenance Technology*

1.4.3 Data Information Sources

- Government Industry Data Exchange Program (GIDEP), GIDEP Operations Center, US Department of the Navy, Corona, California
- Defense Technical Information Center, DTIC-FDAC, 8725 John J. Kingman Road, Suite 0944, Fort Belvoir, Virginia

- Reliability Analysis Center, Rome Air Development Center, Griffis Air Force Base, Rome, New York
- National Technical Information Service, 5285 Port Royal Road, Springfield, Virginia
- Gertman, D. I., Blackman, H. S., *Human Reliability and Safety Analysis Data Handbook*, Wiley, New York, 1994
- American National Standards Institute, 11 W 42nd Street, New York, New York 10036

1.4.4 Standards and Reports

- MIL-STD-785, Reliability Program for Systems and Equipment, Development, and Production, Department of Defense, Washington, DC.
- MIL-HDBK-217, Reliability Prediction of Electronic Equipment, Department of Defense, Washington, DC.
- MIL-STD-721, Definitions of Terms for Reliability and Maintainability, Department of Defense, Washington, DC.
- MIL-STD-1629, Procedures for Performing Failure Mode, Effects, and Criticality Analysis, Department of Defense, Washington, DC.
- MIL-STD-2155, Failure Reporting, Analysis, and Corrective Action, Department of Defense, Washington, DC.
- MIL-STD-756, Reliability Modeling and Prediction, Department of Defense, Washington, DC.
- MIL-STD-882, Systems Safety Program for System and Associated Subsystem and Equipment-Requirements, Department of Defense, Washington, DC.
- DEF-STD-00-55-1, Requirements for Safety-Related Software in Defense Equipment, Department of Defense, Washington, DC.
- MIL-STD-58077, Safety Engineering of Aircraft System, Associated Subsystem and Equipment: General Requirements, Department of Defense, Washington, DC.
- International Electro-Technical Commission (IEC) 61508 SET, Functional Safety of Electrical/Electronic/Programmable Electronic Safety-Related Systems, Parts 1–7, IEC, Geneva, Switzerland, 2000.
- IEC 60950, Safety of Information Technology Equipment, IEC, Geneva, Switzerland, 1999.
- Guide to Reliability-Centered Maintenance, Report No. AMCP 705-2, Department of the Army, Washington, DC, 1985.
- Maintenance Engineering Techniques, Report No. AMCP 706-132, Department of the Army, Washington, DC, 1975.

1.4.5 Books

- Cox, S.J., *Reliability, Safety, and Risk Management: An Integrated Approach*, Butterworth-Heinemann, New York, 1991.
- Keith Mobley, R., Editor in Chief, *Maintenance Engineering Handbook*, McGraw-Hill Education, New York, 2014.
- Dhillon, B.S., *Design Reliability: Fundamentals and Applications*, CRC Press, Boca Raton, Florida, 1999.
- Handley, W., *Industrial Safety Handbook*, McGraw-Hill Book Company, New York, 1969.
- Shooman, M.L., *Probabilistic Reliability: An Engineering Approach*, McGraw-Hill Book Company, New York, 1968.
- Dhillon, B.S., *Engineering Safety: Fundamentals, Techniques, and Applications*, World Scientific Publishing, River Edge, New Jersey, 1996.
- Guy, G.B., Editor, *Reliability on the Move: Safety and Reliability in Transportation*, Elsevier Applied Science, London, 1989.
- Dhillon, B.S., *Transportation Systems Reliability and Safety*, CRC Press, Boca Raton, Florida, 2011.
- Dhillon, B.S., *Robot System Reliability and Safety: A Modern Approach*, CRC Press, Boca Raton, Florida, 2015.
- Dhillon, B.S., *Computer System Reliability: Safety and Usability*, CRC Press, Boca Raton, Florida, 2013.
- Smith, R., *Rules of Thumb for Maintenance and Reliability Engineers*, Elsevier/Butterworth-Heinemann, Amsterdam, Netherlands, 2008.
- Jardine, A. K. S., Tsang, A. H. C., *Maintenance, Replacement, and Reliability: Theory and Applications*, CRC Press, Boca Raton, Florida, 2006.
- Nakagawa, T., *Maintenance Theory of Reliability*, Springer Inc., London, 2005.
- Dhillon, B.S., *Engineering Maintenance: A Modern Approach*, CRC Press, Boca Raton, Florida, 2002.
- Palmer, R. D., *Maintenance Planning and Scheduling Handbook*, McGraw-Hill Book Company, New York, 2012.
- Mobley, R.K., *Maintenance Fundamentals*, Butterworth-Heinemann, Inc., Boston, Massachusetts, 1999.
- August, J., *Applied Reliability-Centered Maintenance*, Penn Well, Tulsa, Oklahoma, 1999.
- Niebel, B.W., *Engineering Maintenance Management*, Marcel Dekker, Inc., New York, 1994.

1.4.6 Conference Proceedings

- *Proceedings of the Maintenance Management Conferences*
- *Proceedings of the European Conferences on Safety and Reliability*
- *Proceedings of the International Conferences on Probabilistic Safety Assessment and Management*
- *Proceedings of the System Safety Conferences*
- *Proceedings of the Annual Reliability and Maintainability Symposium*
- *Proceedings of the ISSAT International Conferences on Reliability and Quality in Design*

1.5 Scope of the Book

Nowadays, engineering systems are an important element of world economy, and each year, a vast sum of money is spent to develop, manufacture, operate, and maintain various types of engineering systems around the globe. Global competition and other factors are forcing manufacturers to produce highly reliable safe and maintainable engineering systems/products. Over the years, a large number of journal and conference proceeding articles, technical reports, etc., on the reliability, the safety, and the maintenance of engineering systems have appeared in the literature. However, to the best of the author's knowledge, there is no book that covers the topics of reliability, safety, and maintenance within its framework. This is a significant impediment to information seekers on these topics, because they have to consult various sources.

Thus, the main objectives of this book are (a) to eliminate the need for professionals and others concerned with engineering system reliability, safety, and maintenance to consult various different and diverse sources in obtaining desired information and (b) to provide up-to-date information on the subject. The book will be useful to many individuals including reliability, safety, and maintenance professionals concerned with engineering systems, engineering system administrators, engineering undergraduate and graduate students, researchers and instructors in the area of engineering systems, and engineers at large.

PROBLEMS

1. Write an essay on engineering system reliability, safety, and maintenance.
2. List seven important facts and figures concerning engineering system reliability, safety, and maintenance.

3. Define the following terms:
   a. *Reliability*
   b. *Maintenance*
   c. *Safety*

4. List four examples concerning engineering system reliability/safety/maintenance-related problems.

5. What is the difference between reliability and availability concerning engineering systems?

6. List six important organizations for obtaining information related to engineering system reliability, safety, and maintenance.

7. List at least four data information sources.

8. What is the difference between the terms *preventive maintenance* and *corrective maintenance*?

9. Define the following terms:
   a. *Continuous task*
   b. *Safety process*
   c. *Predictive maintenance*

10. List six of the most important journals/magazines for obtaining reliability/safety/maintenance-related information.

References

1. Lyman, W.J., Fundamental consideration in preparing a master system plan, *Electrical World*, Vol. 101, 1933, pp. 778–792.
2. Smith, S.A., Service reliability measured by probabilities of outage, *Electrical World*, Vol. 103, 1934, pp. 371–374.
3. Dhillon, B.S., *Reliability and Quality Control: Bibliography on General and Specialized Areas*, Beta Publishers, Gloucester, Ontario, 1992.
4. Goetsch, D.L., *Occupational Safety and Health*, Prentice Hall, Englewood Cliffs, NJ, 1996.
5. Hammer, W., Price, D., *Occupational Safety Management and Engineering*, Prentice Hall, Upper Saddle River, NJ, 2001.
6. Dhillon, B.S., *Engineering Safety: Fundamentals, Techniques, and Applications*, World Scientific Publishing, River Edge, NJ, 2003.
7. *The Volume Library: A Modern Authoritative Reference for Home and School Use*, The South-Western Company, Nashville, TN, 1993.
8. *Factory*, McGraw-Hill, New York, 1882–1968.
9. Kirkman, M.M., *Maintenance of Railways*, C.N. Trivess Printers, Chicago, 1886.
10. Latino, C.J., Hidden treasure: Eliminating chronic failures can cut maintenance costs up to 60%, Report, Reliability Center, Hopewell, VA, 1999.

11. Kane, R.D., Corrosion in petroleum refining and petrochemical operations, in *Metals Handbook*, Vol. 13C: Environments and Industries, edited by S.O. Cramer and B.S. Covino, ASM International, Metals Park, OH, 2003, pp. 967–1014.
12. Backtrom, T., Dooes, M.A., A comparative study of occupational accidents in industries with advanced manufacturing technology, *International Journal of Human Factors in Manufacturing*, Vol. 5, 1995, pp. 267–282.
13. Herrmann, D.S., *Software Safety and Reliability*, IEEE Computer Society Press, Los Alamitos, CA, 1999.
14. Kletz, T., Reflections on safety, *Safety Systems*, Vol. 6, No. 3, 1997, pp. 1–3.
15. Johnston, M., *The Last Nine Minutes: The Story of Flight 981*, Morrow, New York, 1976.
16. Report No. RAR-03/02, *Derailment of Amtrak Auto Train P052-18 on the CSXT Railroad near Crescent City, Florida*, National Transportation Safety Board (NTSB), Washington, DC, 2003.
17. Report No. AAR92-06, *Aircraft Accident Report: United Airlines Flight 585*, National Transportation Safety Board (NTSB), Washington, DC, 1992.
18. Dhillon, B.S., Reliability technology in healthcare systems, *Proceedings of the IASTED International Symposium on Computers, Advanced Technology in Medicine, and Health Care Bioengineering*, 1990, pp. 84–87.
19. National Institute of Standards and Technology (NIST), 100 Bureau Drive, Stop 1070, Gaithersburg, MD, 2002.
20. Hafner, K., Lyon, M., *Where Wizards Stay Up Late: The Origin of the Internet*, Simon and Schuster, New York, 1996.
21. Dhillon, B.S., *Applied Reliability and Quality: Fundamentals, Methods, and Procedures*, Springer, London, 2007.
22. Charette, R.N., *Software Engineering Environments*, Intertext Publications, New York, 1986.
23. Fairley, R.E., *Software Engineering Concepts*, McGraw-Hill, New York, 1985.
24. Nagamachi, M., Ten fatal accidents due to robots in Japan, in *Ergonomics of Hybrid Automated Systems*, edited by W. Karwowski et al., Elsevier, Amsterdam, Netherlands, 1988, pp. 391–396.
25. Clark, D.R., Lehto, M.R., Reliability, maintenance, and safety of robots, in *Handbook of Industrial Robotics*, edited by S.Y. Nof, Wiley, New York, 1999, pp. 717–753.
26. Backtrom, T., Dooes, M., A comparative study of occupational accidents in industries with advanced manufacturing technology, *International Journal of Human Factors in Manufacturing*, Vol. 5, 1995, pp. 267–292.
27. Marx, D.A., *Learning from Our Mistakes: A Review of Maintenance Error Investigation and Analysis Systems* (with Recommendations to FAA), Federal Aviation Administration (FAA), Washington, DC, January 1998.
28. *Human Factors in Airline Maintenance: A Study of Incident Reports*, Bureau of Air Safety Inspection, Department of Transport and Regional Development, Canberra, Australia, 1997.
29. Russell, P.D., Management strategies for accident prevention, *Air Asia*, Vol. 6, 1994, pp. 31–41.
30. Trager, T.A., *Case Study Report on Loss of Safety System Function Events*, Report No. AEOD/C 504, United States Nuclear Regulatory Commission (NRC), Washington, DC, 1985.

31. Daniels, R.W., The formula for improved plant maintainability must include human factors, *Proceedings of the IEEE Conference on Human Factors and Nuclear Safety*, 1985, pp. 242–244.
32. Varma, V., Maintenance training reduces human errors, *Power Engineering*, Vol. 100, 1996, pp. 44–47.
33. Pyy, P., An analysis of maintenance failures at a nuclear power plant, *Reliability Engineering and System Safety*, Vol. 72, 2001, pp. 293–302.
34. Pyy, P., Laakso, K., Reiman, L., A study of human errors related to NPP maintenance activities, *Proceedings of the IEEE Sixth Annual Human Factors Meeting*, 1997, pp. 12.23–12.28.
35. Schatz, W., Fed facts, *Datamation*, Vol. 15, August 1986, pp. 72–73.
36. Kraus, D.C., Gramopadhys, A.K., Effect of team training on aircraft maintenance technicians: Computer-based training versus instructor-based training, *International Journal of Industrial Ergonomics*, Vol. 27, 2001, pp. 141–157.
37. Phillips, E.H., Focus on accident prevention key to future airline safety, *Aviation Week and Space Technology*, Issue No. 5, 1994, pp. 52–53.
38. Christensen, J.M., Howard, J.M., Field experience in maintenance, in *Human Detection and Diagnosis of System Failures*, edited by W. Karwowski et al., Elsevier, Amsterdam, Netherlands, 1988, pp. 391–396.
39. De Rosa, M., Equipment fires cause injuries: NIOSH study reveals trends for equipment fires at U.S. coal mines, *Coal Age*, October 2004, pp. 28–31.
40. Burgess-Limerick, R., Steiner, L., Preventing injuries: Analysis of injuries highlights high priority hazards associated with underground coal mining equipment, *American Longwall Magazine*, August 2006, pp. 19–20.
41. *Definitions of Effectiveness Terms for Reliability, Maintainability, Human Factors, and Safety*, MIL-STD-721, Department of Defense, Washington, DC.
42. McKenna, T., Oliverson, R., *Glossary of Reliability and Maintenance Terms*, Gulf Publishing Company, Houston, TX, 1997.
43. Omdahl, T.P., Editor, *Reliability, Availability, Maintainability (RAM) Dictionary*, ASQC Quality Press, Milwaukee, WI, 1988.
44. *Dictionary of Terms Used in the Safety Profession*, American Society of Safety Engineers (ASSE), 3rd Edition, Des Plaines, IL, 1988.
45. Dhillon, B.S., *Human Reliability: With Human Factors*, Pergamon Press, New York, 1986.

2

Reliability, Safety, and Maintenance Mathematics

2.1 Introduction

Just like the development in other areas of science and engineering, mathematics has also played an important role in the development of reliability, safety, and maintenance fields. The history of mathematics may be traced back to our currently used number symbols often, in the published literature, referred to as the *Hindu–Arabic numeral system* [1]. The first evidence of the use of these symbols is found on stone columns erected by the Scythian emperor of India named Asoka, in around 250 BC [1].

The earliest reference to the probability concept may be traced back to a gambler's manual written by Girolamo Cardano (1501–1576) [2]. However, Pierre Fermat (1601–1665) and Blaise Pascal (1623–1662) were the first two individuals who solved correctly and independently the problem of dividing the winnings in a game of chance [1,2]. Pierre Fermat also introduced the idea of *modern differentiation*. Boolean algebra, which plays an important role in probability theory, is named after George Boole (1815–1864), an English mathematician, who published a pamphlet entitled *The Mathematical Analysis of Logic: Being an Essay Towards a Calculus of Deductive Reasoning* in 1847 [1–3].

Needless to say, a more detailed history of probability and mathematics is available in the studies by Eves [1] and Owen [2]. This chapter presents basic mathematical concepts considered useful to understand the subsequent chapters of this book.

2.2 Median, Arithmetic Mean, and Mean Deviation

A set of engineering system reliability-, safety-, or maintenance-related data is useful only if it is analyzed in an effective manner. More clearly, there are certain characteristics of the data that help to describe the nature of a given set of data, thus making better associated decisions. Thus, this section presents three statistical measures considered quite useful for studying engineering system reliability-, safety-, and maintenance-related data [4,5].

2.2.1 Median

The median is the very middle value or the average of two middle values of a set of data values arranged in an array (i.e., in order of magnitude).

Example 2.1

Assume that the following set of numbers represents engineering system failures occurring over a 13-month period in a manufacturing organization: 40, 75, 20, 25, 10, 30, 15, 35, 5, 18, 36, 24, and 12.

Find the set median.

By arranging the given data values in an array (i.e., in order of magnitude), we get 5, 10, 12, 15, 18, 20, 24, 25, 30, 35, 36, 40, and 75.

Thus, the middle value of the previous set of numbers is 24. Thus, it is the set median.

2.2.2 Arithmetic Mean

Often, arithmetic mean is simply referred to as mean and is expressed by

$$m = \frac{\sum_{i=l}^{n} x_i}{n}, \qquad (2.1)$$

where

n is the total number of data values.
x_i is the data value i, for $i = 1,2,3, ..., n$.
m is the mean value (i.e., arithmetic mean).

Example 2.2

Assume that the quality control department of an engineering system manufacturing company inspected seven identical engineering systems and found 2, 5, 1, 4, 7, 6, and 8 defects in each respective engineering system. Calculate the average number of defects (i.e., arithmetic mean) per engineering system. By substituting the given data values into Equation 2.1, we get

$$m = \frac{2+5+1+4+7+6+8}{7} = 4.71.$$

Thus, the average number of defects per engineering system is 4.71. More specifically, the arithmetic mean of the given data set is 4.71.

2.2.3 Mean Deviation

Mean deviation is a measure of dispersion whose value indicates the degree to which a given set of data tends to spread about a mean value. Mean deviation is defined by

$$MD = \frac{\sum_{i=1}^{n} |x_i - m|}{n}, \qquad (2.2)$$

where
n is the number of data points in a given set of data.
x_i is the data value i, for $i = 1,2,3, ..., n$.
m is the mean value of the given data set.
$|x_1 - m|$ is the absolute value of the deviation of x_i from m.
MD is the mean deviation.

Example 2.3

Calculate the value of the mean deviation of the data set given in Example 2.2.

In Example 2.2, the calculated mean value (i.e., arithmetic mean) of the given data set is $m = 4.71$ defects/engineering system. Thus, using the given data values and this calculated value in Equation 2.2, we get

$$MD = \frac{|2 - 4.71| + |5 - 4.71| + |1 - 4.71| + |4 - 4.71| + |7 - 4.71| + |6 - 4.71| + |8 - 4.71|}{7}$$

$$= 2.04.$$

Thus, the mean deviation of the given data set is 2.04.

2.3 Boolean Algebra Laws

Boolean algebra is used to a degree in various engineering system reliability-, safety-, and maintenance-related studies and is named after its founder, George Boole (1815–1864). Some of its laws considered quite useful to understand subsequent chapters are presented in the following [3–7].

$$M + N = N + M, \tag{2.3}$$

where
  M is an arbitrary set or event.
  N is an arbitrary set or event.
  $+$ denotes the union of sets or events.

$$M \cdot N = N \cdot M, \tag{2.4}$$

where the dot between M and N or N and M denotes the intersection of sets or events. It is to be noted that many times, Equation 2.4 is written without a dot (e.g., MN), but it still conveys the same meaning.

$$MM = M, \tag{2.5}$$

$$N + N = N, \tag{2.6}$$

$$N + NM = N, \tag{2.7}$$

$$M(M + N) = M, \tag{2.8}$$

$$N(M + P) = NM + NP, \tag{2.9}$$

where
  P is an arbitrary set or event.

$$(M + N)(M + P) = M + NP, \tag{2.10}$$

$$(M + N) + P = M + (N + P), \tag{2.11}$$

$$(MN)P = M(NP). \tag{2.12}$$

It is to be noted that in the published literature, Equations 2.11 and 2.12 are referred to as associative law; Equations 2.9 and 2.10, as distributive law; Equations 2.7 and 2.8, as absorption law; Equations 2.5 and 2.6, as idempotent law; and Equations 2.3 and 2.4, as commutative law [8].

2.4 Probability Definition and Properties

Probability is defined as follows [9]:

$$P(Y) = \lim_{n \to \infty} \left(\frac{N}{n} \right), \tag{2.13}$$

where
  $P(Y)$ is the probability of occurrence of event Y.
  N is the number of times that event Y occurs in the n repeated experiments.

Some of the basic properties of probability are as follows [6,9]:

- The probability of occurrence of an event, say event X, is

$$0 \le P(X) \le 1. \tag{2.14}$$

- The probability of occurrence and nonoccurrence of an event, say event X, is always

$$P(X) + P(\bar{X}) = 1, \tag{2.15}$$

where
  $P(X)$ is the probability of occurrence of event X.
  $P(\bar{X})$ is the probability of nonoccurrence of event X.

- The probability of the union of n mutually exclusive events is given by

$$P(X_1 + X_2 + \ldots + X_n) = \sum_{i=1}^{n} P(X_i), \tag{2.16}$$

where
  $P(X_i)$ is the probability of occurrence of event X_i, for $i = 1,2,3, \ldots, n$.

- The probability of the union of n-independent events is expressed by

$$P(X_1 + X_2 + \ldots - + X_n) = 1 - \prod_{i=1}^{n} \left[1 - P(X_i) \right]. \tag{2.17}$$

- The probability of an interaction of n-independent events is given by

$$P(X_1 X_2 X_3 \ldots X_n) = P(X_1)P(X_2)P(X_3)\ldots P(X_n). \tag{2.18}$$

Example 2.4

Assume that an engineering system is made up of two very critical subsystems, say subsystem X_1 and subsystem X_2. The malfunctioning of either subsystem can result in system failure. The failure probability of subsystems X_1 and X_2 is 0.06 and 0.04, respectively.

Calculate the probability of the engineering system failure if both of these subsystems fail independently.

By substituting the given data values into Equation 2.17, we get

$$
\begin{aligned}
P(X_1 + X_2) &= 1 - \prod_{i=1}^{2}[1 - P(X_i)] \\
&= P(X_1) + P(X_2) - P(X_1)P(X_2) \\
&= 0.06 + 0.04 - (0.06)(0.04) \\
&= 0.0976.
\end{aligned}
$$

Thus, the probability of the engineering system failure is 0.0976.

2.5 Useful Mathematical Definitions

This section presents a number of mathematical definitions considered useful for performing various types of engineering system reliability, safety, and maintenance studies.

2.5.1 Cumulative Distribution Function

For a continuous random variable, the cumulative distribution function is defined by Dhillon [8] and Mann et al. [9].

$$F(t) = \int_{-\infty}^{t} f(x)\,dx, \tag{2.19}$$

where
x is a continuous random variable.
$f(x)$ is the probability density function.
$F(t)$ is the cumulative distribution function.

For $t = \infty$, Equation 2.19 becomes

$$F(\infty) = \int_{-\infty}^{\infty} f(x)\,dx$$

$$= 1.$$

(2.20)

It simply means that the total area under a probability density curve is equal to unity. Usually, in reliability, safety, and maintenance studies of engineering systems, Equation 2.19 is simply written as

$$F(t) = \int_{0}^{t} f(x)\,dx.$$

(2.21)

Example 2.5

Assume that the probability (i.e., failure) density function of an engineering system is

$$f(t) = \lambda_{es} e^{-\lambda_{es} t} \quad \text{for } t \geq 0, \ \lambda_{es} > 0,$$

(2.22)

where
λ_{es} is the engineering system failure rate.
t is the time (i.e., a continuous random variable).
$f(t)$ is the probability density function (usually, in the area of reliability engineering, it is known as the *failure density function*).

Obtain an expression for the engineering system cumulative distribution function.
By substituting Equation 2.22 into Equation 2.19, we obtain

$$F(t) = \int_{0}^{t} \lambda_{es} e^{-\lambda_{es} t}\,dt$$

$$= 1 - e^{-\lambda_{es} t}.$$

(2.23)

Thus, Equation 2.23 is the expression for the engineering system cumulative distribution function.

2.5.2 Probability Density Function

For a continuous random variable, the probability density function is expressed by Mann et al. [9] as

$$f(t) = \frac{dF(t)}{dt},$$

(2.24)

where

F(t) is the cumulative distribution function.

f(t) is the probability density function.

Example 2.6

Prove with the aid of Equation 2.23 that Equation 2.22 is the probability density function.

By inserting Equation 2.23 into Equation 2.24, we obtain

$$f(t) = \frac{d\left(1 - e^{-\lambda_{es}t}\right)}{dt}$$

$$= \lambda_{es}e^{-\lambda_{es}t}. \tag{2.25}$$

Equations 2.25 and 2.22 are identical.

2.5.3 Expected Value

The expected value of a continuous random variable is expressed by Mann et al. [9] as

$$E(t) = \int_{-\infty}^{\infty} tf(t)\,dt, \tag{2.26}$$

where

E(t) is the expected value (i.e., mean value) of the continuous random variable t.

Example 2.7

Find the expected value (i.e., mean value) of the probability (failure) density function defined by Equation 2.22.

By inserting Equation 2.22 into Equation 2.26, we obtain

$$E(t) = \int_{0}^{\infty} t\lambda_{es}e^{-\lambda_{es}t}\,dt$$

$$= \left[-te^{-\lambda_{es}t}\right]_{0}^{\infty} - \left[\frac{-e^{-\lambda_{es}t}}{\lambda_{es}}\right]_{0}^{\infty} \tag{2.27}$$

$$= \frac{1}{\lambda_{es}}.$$

Thus, the expected value (i.e., mean value) of the probability (failure) density function defined by Equation 2.22 is given by Equation 2.27.

2.5.4 Laplace Transform

Laplace transform is named after a French mathematician, Pierre-Simon Laplace (1749–1827), and is defined by Eves [1], Spiegel [10], Oberhettinger and Badic [11] as

$$f(s) = \int_0^\infty f(t)e^{-st} \, dt, \tag{2.28}$$

where
  t is a variable.
  s is the Laplace transform variable.
  $f(s)$ is the Laplace transform of function $f(t)$.

An example of obtaining Laplace transform using Equation 2.28 is presented in the following, and Laplace transforms of some frequently occurring functions in reliability, safety, or maintenance areas are presented in Table 2.1 [10,11].

TABLE 2.1

Laplace Transforms of Some Functions

No.	$f(t)$	$f(s)$
1	C (a constant)	$\dfrac{C}{s}$
2	$e^{-\theta t}$	$\dfrac{1}{s+\theta}$
3	t^m for $m = 0, 1, 2, 3,$	$\dfrac{m!}{s^{m+1}}$
4	$te^{-\theta t}$	$\dfrac{1}{(s+\theta)^2}$
5	$\alpha_1 f_1(t) + \alpha_2 f_2(t)$	$\alpha_1 f_1(s) + \alpha_2 f_2(s)$
6	$\dfrac{df(t)}{dt}$	$sf(s) - f(0)$
7	$tf(t)$	$-\dfrac{df(s)}{ds}$

Example 2.8

Obtain the Laplace transform of the following function:

$$f(t) = e^{-\theta t},\tag{2.29}$$

where
   θ is a constant.

By inserting Equation 2.29 into Equation 2.28, we get

$$f(s) = \int_0^\infty e^{-\theta t} e^{-st}\, dt$$

$$= -\frac{e^{-(s+\theta)t}}{s+\theta}\bigg|_0^\infty\tag{2.30}$$

$$= \frac{1}{s+\theta}.$$

Thus, Equation 2.30 is the Laplace transform of Equation 2.29.

2.5.5 Final Value Theorem Laplace Transform

If the following limits exist, then the final value theorem may be stated as

$$\lim_{t\to\infty} f(t) = \lim_{s\to 0} sf(s).\tag{2.31}$$

Example 2.9

Prove using the following equation that the left-hand side of Equation 2.31 is equal to its right-hand side:

$$f(t) = \frac{\alpha}{(\theta+\alpha)} + \frac{\theta}{(\theta+\alpha)} e^{-(\theta+\alpha)t},\tag{2.32}$$

where
   θ and α are constants.

By substituting Equation 2.32 into the left-hand side of Equation 2.31, we obtain

$$\lim_{t\to\infty}\left[\frac{\alpha}{(\theta+\alpha)} + \frac{\theta}{(\theta+\alpha)} e^{-(\theta+\alpha)t}\right] = \frac{\alpha}{(\theta+\alpha)}.\tag{2.33}$$

Using Table 2.1, we obtain the following Laplace transforms of Equation 2.32:

$$f(s) = \frac{\alpha}{s(\theta + \alpha)} + \frac{\theta}{(\theta + \alpha)} \cdot \frac{1}{(s + \theta + \alpha)}. \tag{2.34}$$

By substituting Equation 2.34 into the right-hand side of Equation 2.31, we get

$$\lim_{s \to 0} \left[\frac{s\alpha}{s(\theta + \alpha)} + \frac{s\theta}{(\theta + \alpha)(s + \theta + \alpha)} \right] = \frac{\alpha}{(\theta + \alpha)}. \tag{2.35}$$

As the right-hand sides of Equations 2.33 and 2.35 are the same, this proves that the left-hand side of Equation 2.31 is equal to its right-hand side.

2.6 Solving First-Order Differential Equations with Laplace Transforms

Usually, Laplace transforms are used for finding solutions to first-order linear differential equations involved in reliability-, safety-, and maintenance analysis-related studies of engineering systems. The following example demonstrates the finding of solutions to a set of linear first-order differential equations, describing an engineering system with respect to reliability and safety.

Example 2.10

Assume that an engineering system can be in any of the three states: operating normally, failed safely, or failed unsafely. The following three first-order linear differential equations describe the engineering system under consideration:

$$\frac{dP_0(t)}{dt} + \left(\lambda_s + \lambda_u \right) P_0(t) = 0, \tag{2.36}$$

$$\frac{dP_1(t)}{dt} - \lambda_s P_0(t) = 0, \tag{2.37}$$

$$\frac{dP_2(t)}{dt} - \lambda_u P_0(t) = 0, \tag{2.38}$$

where

$P_{(t)}$ is the probability that the engineering system is in state i at time t, for $i = 0$ (operating normally), $i = 1$ (failed safely), and $i = 2$ (failed unsafely).

λ_s is the engineering system constant safe failure rate.

λ_u is the engineering system constant unsafe failure rate.

At time $t = 0$, $P_0(0) = 1$, $P_1(0) = 0$, and $P_2(0) = 0$.

Solve differential Equations 2.36 through 2.38 by using Laplace transforms.

Using Table 2.1, Equations 2.36 through 2.38, and the specified initial conditions, we get

$$sP_0(s) - 1 + (\lambda_s + \lambda_u)P_0(s) = 0, \tag{2.39}$$

$$sP_1(s) - \lambda_s P_0(s) = 0, \tag{2.40}$$

$$sP_2(s) - \lambda_u P_0(s) = 0. \tag{2.41}$$

By solving Equations 2.39 through 2.41, we get

$$P_0(s) = \frac{1}{\left(s + \lambda_s + \lambda_u\right)}, \tag{2.42}$$

$$P_1(s) = \frac{\lambda_s}{s\left(s + \lambda_s + \lambda_u\right)}, \tag{2.43}$$

$$P_2(s) = \frac{\lambda_u}{s\left(s + \lambda_s + \lambda_u\right)}. \tag{2.44}$$

By taking the inverse Laplace transforms of Equations 2.42 through 2.44, we obtain

$$P_0(t) = e^{-(\lambda_s + \lambda_u)t}, \tag{2.45}$$

$$P_1(t) = \frac{\lambda_s}{(\lambda_s + \lambda_u)}\left[1 - e^{-(\lambda_s + \lambda_u)t}\right], \tag{2.46}$$

$$P_2(t) = \frac{\lambda_u}{(\lambda_s + \lambda_u)}\left[1 - e^{-(\lambda_s + \lambda_u)t}\right]. \tag{2.47}$$

Thus, Equations 2.45 through 2.47 are the solutions to differential Equations 2.36 through 2.38.

2.7 Statistical Distributions

Although there are a large number of statistical/probability distributions, this section presents just five such distributions considered useful for performing various types of engineering system reliability-, safety-, and maintenance-related studies [12–15].

2.7.1 Binomial Distribution

Binomial distribution, a discrete random variable statistical distribution, is used in situations where one is concerned with the probabilities of outcome such as the number of occurrences (e.g., failures) in a sequence of specified number of trials. More clearly, each trial has two possible outcomes (e.g., success or failure), but the probability of each trial remains constant/unchanged. The distribution is also known as the *Bernoulli distribution*, after its founder, Jakob Bernoulli (1654–1705) [1]. The distribution probability density function is defined by

$$f(x) = \frac{m!}{x!(m-x)!}p^x q^{m-x} \quad \text{for } x = 0, 1, 2, 3, \ldots, m, \tag{2.48}$$

where
p is the single trial probability of occurrence (e.g., success).
q is the single trial probability of nonoccurrence (e.g., failure).
x is the number of nonoccurrences (e.g., failures) in a total of m trials.

The cumulative distribution function is expressed by

$$F(x) = \sum_{j=0}^{x} \frac{m!}{j!(m-j)!}p^j q^{m-j}, \tag{2.49}$$

where
$F(x)$ is the probability of x or less nonoccurrences (e.g., failures) in m trials.

2.7.2 Exponential Distribution

Exponential distribution is one of the simplest continuous random variable statistical/probability distributions frequently used in the industrial sector, for performing various types of engineering system reliability-, safety-, and maintenance-related studies. Its probability density function is defined by Dhillon [8,15] and Davis [16]:

$$f(t) = \alpha e^{-\alpha t} \quad \text{for } \alpha > 0, \ t \geq 0, \qquad (2.50)$$

where
  $f(t)$ is the probability density function.
  α is the distribution parameter.
  t is the time (i.e., a continuous random variable).

By inserting Equation 2.50 into Equation 2.21, we get the following expression for the cumulative distribution function:

$$F(t) = 1 - e^{-\alpha t}. \qquad (2.51)$$

With the aid of Equations 2.26 and 2.50, we obtain the following expression for the distribution expected value (i.e., mean value):

$$E(t) = \frac{1}{\alpha}. \qquad (2.52)$$

Example 2.11

Assume that the mean time to failure of an engineering system is 1500 hours. Calculate the probability of failure of the engineering system during a 500-hour mission with the aid of Equations 2.51 and 2.52.
  By inserting the specified data value into Equation 2.52, we obtain

$$\alpha = \frac{1}{1500} = 0.00066 \text{ failures per hour.}$$

By substituting the calculated and the specified data values into Equation 2.51, we get

$$F(500) = 1 - e^{-(0.00066)(500)}$$

$$= 0.2834.$$

Thus, the probability of failure of the engineering system during the 500-hour mission is 0.2834.

2.7.3 Rayleigh Distribution

Rayleigh distribution is a continuous random variable statistical/probability distribution named after its founder, John Rayleigh (1842–1919) [1], and its probability density function is defined by

$$f(t) = \left(\frac{1}{\theta^2}\right) t e^{-\left(\frac{t}{\theta}\right)^2} \quad \text{for } \theta > 0,\ t \geq 0, \tag{2.53}$$

where
  θ is the distribution parameter.
  t is the time (i.e., a continuous random variable).

By substituting Equation 2.53 into Equation 2.21, we obtain the following expression for the cumulative distribution function:

$$F(t) = 1 - e^{-(t/\theta)^2}. \tag{2.54}$$

By inserting Equation 2.53 into Equation 2.26, we get the following expression for the distribution expected value (i.e., mean value):

$$E(t) = \theta\Gamma\left(\frac{3}{2}\right), \tag{2.55}$$

where
  $\Gamma(.)$ is the gamma function and is defined by

$$\Gamma(n) = \int_0^\infty t^{n-1} e^{-t}\, dt, \quad \text{for } n > 0. \tag{2.56}$$

2.7.4 Weibull Distribution

Weibull distribution is a continuous random variable statistical/probability distribution named after its founder, Waloddi Weibull, a Swedish professor in mechanical engineering [17]. The probability density function of the distribution is defined by

$$f(t) = \frac{b t^{b-1}}{\theta^b} e^{-(t/\theta)^b} \quad \text{for } b > 0,\ \theta > 0,\ t \geq 0, \tag{2.57}$$

where
   θ and b are the distribution scale and shape parameters, respectively.
   t is the time (i.e., a continuous random variable).

By inserting Equation 2.57 into Equation 2.21, we obtain the following equation for the cumulative distribution function:

$$F(t) = 1 - e^{-(t/\theta)^b}. \tag{2.58}$$

By substituting Equation 2.57 into Equation 2.26, we obtain the following equation for the distribution expected value (i.e., mean value):

$$E(t) = \theta\Gamma\left(1 + \frac{1}{b}\right). \tag{2.59}$$

It is to be noted that for $b = 1$ and $b = 2$, the exponential and Rayleigh distributions are the special cases of the Weibull distribution, respectively.

2.7.5 Bathtub Hazard Rate Curve Distribution

Bathtub hazard rate curve distribution is a continuous random variable statistical/probability distribution developed in 1981 [18]. In the published literature by authors around the world, it is generally referred to as the *Dhillon distribution/model/law* [19–40]. The distribution can represent bathtub-shaped, decreasing and increasing hazard rates of engineering systems.
   The distribution probability density function is defined by Dhillon [18] as

$$f(t) = b\theta(\theta t)^{b-1} e^{-\left[e^{-(\theta t)^b} - (\theta t)^b - 1\right]}, \quad \text{for } t \geq 0, \theta > 0, b > 0, \tag{2.60}$$

where
   θ and b are the distribution scale and shape parameters, respectively.
   t is the time (i.e., a continuous random variable).

By substituting Equation 2.60 into Equation 2.21, we obtain the following equation for the cumulative distribution function:

$$F(t) = 1 - e^{-\left[e^{(\theta t)^b} - 1\right]}. \tag{2.61}$$

It is to be noted that for $b = 0.5$, this probability distribution gives the bathtub-shaped hazard rate curve, and for $b = 1$, it gives the extreme value distribution. More specifically, at $b = 1$, the extreme value statistical/probability distribution is the special case of this distribution.

PROBLEMS

1. Assume that the quality control department of an engineering system manufacturing company inspected nine identical engineering systems and found 4, 1, 5, 7, 2, 3, 8, 6, and 10 defects in each respective engineering system. Calculate the average number of defects (i.e., arithmetic mean) per engineering system.
2. What is absorption law?
3. Calculate the mean deviation of the data set given in question 1.
4. What are the basic probability properties?
5. Define the following items:
   a. Probability
   b. Cumulative distribution function
6. Define the following items:
   a. Expected value of a continuous random variable
   b. Laplace transform
7. Write down probability density functions for the following distributions:
   a. Exponential distribution
   b. Rayleigh distribution
8. What are the special case distributions of the Weibull distribution?
9. Write down the probability density function for the bathtub hazard rate curve distribution.
10. Prove Equations 2.45 through 2.47 by using Equations 2.42 through 2.44.

References

1. Eves, H., *An Introduction to the History of Mathematics*, Holt, Rinehart & Winston, New York, 1976.
2. Owen, D.B., Editor, *On the History of Statistics and Probability*, Marcel Dekker, New York, 1976.
3. Lipschutz, S., *Set Theory*, McGraw-Hill Book Company, New York, 1964.
4. Spiegel, M.R., *Probability and Statistics*, McGraw-Hill Book Company, New York, 1975.
5. Spiegel, M.R., *Statistics*, McGraw-Hill Book Company, New York, 1961.

6. Lipschutz, S., *Probability*, McGraw-Hill Book Company, New York, 1965.
7. *Fault Tree Handbook*, Report No. NUREG-0492, US Nuclear Regulatory Commission, Washington, DC, 1981.
8. Dhillon, B.S., *Computer System Reliability: Safety and Usability*, CRC Press, Boca Raton, FL, 2013.
9. Mann, N.R., Schafer, R.E., Singpurwalla, N.P., *Methods for Statistical Analysis of Reliability and Life Data*, Wiley, New York, 1974.
10. Spiegel, M.R., *Laplace Transforms*, McGraw-Hill Book Company, New York, 1965.
11. Oberhettinger, F., Badic, L., *Tables of Laplace Transforms*, Springer-Verlag, New York, 1973.
12. Patel, J.K., Kapadia, C.H., Owen, D.H., *Handbook of Statistical Distributions*, Marcel Dekker, New York, 1976.
13. Shooman, M.L., *Probabilistic Reliability: An Engineering Approach*, McGraw-Hill Book Company, New York, 1968.
14. Dhillon, B.S., *Reliability Engineering in Systems Design and Operation*, Van Nostrand Reinhold, New York, 1983.
15. Dhillon, B.S., *Design Reliability: Fundamentals and Applications*, CRC Press, Boca Raton, FL, 1999.
16. Davis, D.J., Analysis of some failure data, *Journal of the American Statistical Association*, 1952, pp. 113–150.
17. Weibull, W., A statistical distribution function of wide applicability, *Journal of Applied Mechanics*, Vol. 18, 1951, pp. 293–297.
18. Dhillon, B.S., Life distributions, *IEEE Transactions on Reliability*, Vol. 30, 1981, pp. 457–460.
19. Baker, R.D., Nonparametric estimation of the renewal function, *Computers Operations Research*, Vol. 20, No. 2, 1993, pp. 167–178.
20. Cabana, A., Cabana, E.M., Goodness-of-fit to the exponential distribution, focused on Weibull alternatives, *Communications in Statistics-Simulation and Computation*, Vol. 34, 2005, pp. 711–723.
21. Grane, A., Fortiana, J., A directional test of exponentiality based on maximum correlations, *Metrika*, Vol. 73, 2011, pp. 255–274.
22. Henze, N., Meintnis, S.G., Recent and classical tests for exponentiality: A partial review with comparisons, *Metrika*, Vol. 61, 2005, pp. 29–45.
23. Jammalamadaka, S.R., Taufer, E., Testing exponentiality by comparing the empirical distribution function of the normalized spacings with that of the original data, *Journal of Nonparametric Statistics*, Vol. 15, No. 6, 2003, pp. 719–729.
24. Hollander, M., Laird, G., Song, K.S., Nonparametric interference for the proportionality function in the random censorship model, *Nonparametric Statistics*, Vol. 15, No. 2, 2003, pp. 151–169.
25. Jammalamadaka, S.R., Taufer, E., Use of mean residual life in testing departures from exponentiality, *Journal of Nonparametric Statistics*, Vol. 18, No. 3, 2006, pp. 277–292.
26. Kunitz, H., Pamme, H., The mixed gamma ageing model in life data analysis, *Statistical Papers*, Vol. 34, 1993, pp. 303–318.
27. Kunitz, H., A new class of bathtub-shaped hazard rates and its application in comparison of two test-statistics, *IEEE Transactions on Reliability*, Vol. 38, No. 3, 1989, pp. 351–354.
28. Meintanis, S.G., A class of tests for exponentiality based on a continuum of moment conditions, *Kybernetika*, Vol. 45, No. 6, 2009, pp. 946–959.

29. Morris, K., Szynal, D., Goodness-of-fit tests based on characterizations involving moments of order statistics, *International Journal of Pure and Applied Mathematics*, Vol. 38, No. 1, 2007, pp. 83–121.
30. Na., M.H., Spline hazard rate estimation using censored data, *Journal of KSIAM*, Vol. 3, No. 2, 1999, pp. 99–106.
31. Morris, K., Szynal, D., Some U-Statistics in goodness-of-fit tests derived from characterizations via record values, *International Journal of Pure and Applied Mathematics*, Vol. 46, No. 4, 2008, pp. 339–414.
32. Nam, K.H., Park, D.H., Failure rate for Dhillon model, *Proceedings of the Spring Conference of the Korean Statistical Society*, 1997, pp. 114–118.
33. Nimoto, N., Zitikis, R., The Atkinson index, the Moran statistic, and testing exponentiality, *Journal of the Japan Statistical Society*, Vol. 38, No. 2, 2008, pp. 187–205.
34. Nam, K.H., Chang, S.J., Approximation of the renewal function for Hjorth model and Dhillon model, *Journal of the Korean Society for Quality Management*, Vol. 34, No. 1, 2006, pp. 34–39.
35. Noughabi, H.A., Arghami, N.R., Testing exponentiality based on characterizations of the exponential distribution, *Journal of Statistical Computation and Simulation*, Vol. 1, 2011, pp. 1–11.
36. Szynal, D., *Goodness-of-Fit Tests Derived from Characterizations of Continuous Distributions, Stability in Probability*, Banach Center Publications, Institute of Mathematics, Polish Academy of Sciences, Warszawa, Poland, 2010, pp. 203–223.
37. Szynal, D., Wolynski, W., Goodness-of-fit tests for exponentiality and Rayleigh distribution, *International Journal of Pure and Applied Mathematics*, Vol. 78, No. 5, 2012, pp. 751–772.
38. Nam, K.H., Park, D.H., A study on trend changes for certain parametric families, *Journal of the Korean Society for Quality Management*, Vol. 23, No. 3, 1995, pp. 93–101.
39. Srivastava, A.K., Validation analysis of Dhillon model on different real data sets for reliability modelling, *International Journal of Advance Foundation and Research in Computer (IJAFRC)*, Vol. 1, Issue 9, September 2014, pp. 18–31.
40. Srivastava, A.K., Kumar, V., A study of several issues of reliability modelling for a real dataset using different software reliability models, *International Journal of Emerging Technology and Advanced Engineering*, Vol. 5, No. 12, 2015, pp. 49–57.

3

Reliability, Safety, and Maintenance Basics

3.1 Introduction

Nowadays, the reliability of engineering systems has become a very important issue during the design process due to the increasing dependence of our daily lives and schedules on the satisfactory functioning of these systems. Some examples of these systems are computers, automobiles, aircraft, nuclear power-generating reactors, and space satellites. Over the years, many methods and approaches have been developed to improve the reliability of engineering systems at large.

Nowadays, engineering systems have become highly complex and sophisticated. The safety of these systems has become a challenging issue. Over the years, various types of approaches and methods have been used to improve the safety of engineering systems.

Since the Industrial Revolution, the maintenance of engineering systems/equipment in the field has been a challenging issue. Although, over the years, impressive progress has been made in maintaining engineering systems/equipment in the field in an effective manner, the maintenance of engineering systems/equipment is still a challenging issue due to factors such as complexity, competition, and cost. Needless to say, over the years, various types of methods and approaches have been developed for improving maintenance of engineering systems/equipment in the field.

This chapter presents various reliability, safety, and maintenance basics considered useful to understand the subsequent chapters of this book.

3.2 Bathtub Hazard Rate Curve

The bathtub hazard rate curve shown in Figure 3.1 is usually used for describing the failure rate of engineering systems/equipment. The curve is called the *bathtub hazard rate curve* because it resembles the shape of a bathtub.

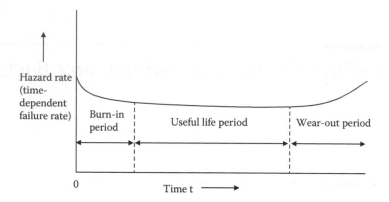

FIGURE 3.1
Bathtub hazard rate curve.

As shown in Figure 3.1, the curve is divided into three sections: burn-in period, useful life period, and wear-out period.

During the burn-in period, the system/item hazard rate decreases with time, and some of the reasons for the occurrence of failures during this time period are inadequate debugging, poor manufacturing methods and processes, poor quality control, human error, and substandard materials and workmanship [1,2]. Three other terms used in the published literature for this decreasing hazard rate region are *debugging region*, *infant mortality region*, and *break-in region*.

During the useful life period, the hazard rate remains constant. Some of the reasons for the occurrence of failures in this region are as follows [1,2]:

- Higher random stress than expected
- Low safety factors
- Undetectable defects
- Abuse
- Natural failures
- Human errors

Finally, during the wear-out period, the hazard rate increases with time t. Some of the reasons for the occurrence of failures in this region are wear from aging, wrong overhaul practices, wear due to friction, corrosion, and creep; short designed-in life of the item/system under consideration; and poor maintenance practices [1,2].

Mathematically, the following equation can be used for representing Figure 3.1 bathtub hazard rate curve [3]:

$$\lambda(t) = \alpha\beta(\alpha t)^{\beta-1} e^{(\alpha t)^{\beta}}, \tag{3.1}$$

where
  β is the shape parameter.
  α is the scale parameter.
  t is time.
  $\lambda(t)$ is the hazard rate (time-dependent failure rate).

At β = 0.5, Equation 3.1 gives the shape of the bathtub hazard rate curve shown in Figure 3.1.

3.3 General Reliability Formulas

A number of general reliability formulas are often used in performing various types of reliability analysis. Four of these formulas are presented in the following.

3.3.1 Probability (or Failure) Density Function

The probability (or failure) density function is expressed by Dhillon [2]

$$f(t) = -\frac{dR(t)}{dt},\qquad(3.2)$$

where
  $R(t)$ is the system/item reliability at time *t*.
  $f(t)$ is the probability (or failure) density function.

Example 3.1

Assume that an engineering system's reliability is expressed by

$$R_{es}(t) = e^{-\lambda_{es}t},\qquad(3.3)$$

where
  $R_{es}(t)$ is the engineering system reliability at time *t*.
  λ_{es} is the engineering system constant failure rate.

Obtain an expression for the probability (or failure) density function of the engineering system by using Equation 3.2.
  By substituting Equation 3.3 into Equation 3.2, we obtain

$$f(t) = -\frac{de^{-\lambda_{es}t}}{dt}$$

$$= \lambda_{es}e^{-\lambda_{es}t}.\qquad(3.4)$$

Thus, Equation 3.4 is the expression for the probability (or failure) density function of the engineering system.

3.3.2 Hazard Rate (or Time-Dependent Failure Rate) Function

The hazard rate (or time-dependent failure rate) function is expressed by

$$\lambda(t) = \frac{f(t)}{R(t)}, \tag{3.5}$$

where

$\lambda(t)$ is the item/system hazard rate (or time-dependent failure rate) function.

Inserting Equation 3.2 into Equation 3.5 yields

$$\lambda(t) = -\frac{1}{R(t)} \cdot \frac{dR(t)}{dt}. \tag{3.6}$$

Example 3.2

Obtain an expression for the hazard rate of the engineering system by using Equations 3.3 and 3.6 and comment on the final result.

By inserting Equation 3.3 into Equation 3.6, we obtain

$$\lambda(t) = -\frac{1}{e^{-\lambda_{es}t}} \cdot \frac{de^{-\lambda_{es}t}}{dt} \tag{3.7}$$

$$= \lambda_{es}.$$

Thus, the hazard rate of the engineering system is given by Equation 3.7. The right-hand side of this equation is not a function of time t. In other words, it is constant. Usually, it is referred to as the *constant failure rate* of an item/system (in this case, of the engineering system) because it does not depend on time t.

3.3.3 General Reliability Function

The general reliability function can be obtained by using Equation 3.6. Thus, by rearranging Equation 3.6, we get

$$-\lambda(t)dt = \frac{1}{R(t)} \cdot dRt. \tag{3.8}$$

By integrating both sides of Equation 3.8 over the time interval $[0, t]$, we obtain

$$-\int_0^t \lambda(t)\,dt = \int_1^{R(t)} \frac{1}{R(t)}\,dR(t). \tag{3.9}$$

Since, at $t = 0$, $R(t) = 1$.

By evaluating the right-hand side of Equation 3.9 and then rearranging, we get

$$\ln R(t) = -\int_0^t \lambda(t)\,dt. \tag{3.10}$$

Thus, from Equation 3.10, we obtain

$$R(t) = e^{-\int_0^t \lambda(t)\,dt}. \tag{3.11}$$

Equation 3.11 is the general expression for the reliability function. It can be used to obtain the reliability expression of an item/system when its times to failure follow any time-continuous probability distribution (e.g., Rayleigh, Weibull, and exponential).

Example 3.3

Assume that the times to failure of an engineering system are exponentially distributed. Thus, its failure rate is constant and is 0.004 failures per hour. Calculate the reliability of the engineering system for an 8-hour mission.

By inserting the specified data values into Equation 3.11, we obtain

$$R(8) = e^{-\int_0^8 (0.004)\,dt}$$

$$= e^{-(0.004)(8)}$$

$$= 0.9685.$$

Thus, the reliability of the engineering system is 0.9685. In other words, there is 96.85% chance that the engineering system will not malfunction during the stated period.

3.3.4 Mean Time to Failure

The mean time to failure (MTTF) can be obtained by using any of the following three formulas [4,5]:

$$MTTF = E(t) = \int_0^\infty t f(t) dt \tag{3.12}$$

or

$$MTTF = \lim_{s \to 0} R(s) \tag{3.13}$$

or

$$MTTF = \int_0^\infty R(t) dt, \tag{3.14}$$

where
   $MTTF$ is the mean time to failure.
   $E(t)$ is the expected value.
   s is the Laplace transform variable.
   $R(s)$ is the Laplace transform of the reliability function $R(t)$.

Example 3.4

Prove with the aid of Equation 3.3 that Equations 3.13 and 3.14 yield the identical result for the engineering system mean time to failure.
   By taking the Laplace transform of Equation 3.3, we obtain

$$R_{es}(s) = \int_0^\infty e^{-st} e^{-\lambda_{es} t} dt$$
$$= \frac{1}{s + \lambda_{es}}, \tag{3.15}$$

where
   $R_{es}(s)$ is the Laplace transform of the engineering system reliability function $R_{es}(t)$.

By substituting Equation 3.15 into Equation 3.13, we get

$$MTTF = \lim_{s \to 0} \frac{1}{s + \lambda_{es}}$$
$$= \frac{1}{\lambda_{es}}. \tag{3.16}$$

By inserting Equation 3.3 into Equation 3.14, we get

$$MTTF = \int_0^{\infty} e^{-\lambda_{es} t}\, dt$$
$$= \frac{1}{\lambda_{es}}. \tag{3.17}$$

Equations 3.16 and 3.17 are the same. It proves that Equations 3.13 and 3.14 yield the identical result for the engineering system mean time to failure.

3.4 Reliability Configurations

An engineering system can form various configurations in performing reliability analysis. This section is concerned with the reliability evaluation of such commonly occurring configurations.

3.4.1 Series Configuration

The series configuration is the simplest reliability configuration/network, and its block diagram is shown in Figure 3.2. The diagram denotes an n unit series system, and each block in the diagram represents a unit. For the successful operation of the series system, all its n units must operate normally. In other words, if any one of the n units fails, the series system/configuration/ network fails.

The series system/configuration/network reliability, shown in Figure 3.2, is expressed by

$$R_{ss} = P(E_1 E_2 E_3 \ldots E_n), \tag{3.18}$$

where
R_{ss} is the series system reliability.
E_j is the successful operation (i.e., success event) of unit j, for $j = 1, 2, 3, \ldots, n$.
$P(E_1 E_2 E_3 \ldots E_n)$ is the probability of occurrence of events $E_1, E_2, E_3, \ldots, E_n$.

FIGURE 3.2
An n unit series system (configuration).

For independently failing units, Equation 3.18 becomes

$$R_{ss} = P(E_1)P(E_2)P(E_3)\dots P(E_n), \tag{3.19}$$

where
$P(E_j)$ is the occurrence probability of event E_j, for $j = 1, 2, 3, \dots, n$.

If we let $R_j = P(E_j)$, for $j = 1, 2, 3, \dots, n$, Equation 3.19 becomes

$$R_{ss} = R_1 R_2 R_3 \dots R_n$$

$$= \prod_{j=1}^{n} R_j, \tag{3.20}$$

where
R_j is the reliability of unit j, for $j = 1, 2, 3, \dots, n$.

For constant failure rate λ_j of unit j from Equation 3.11 (i.e., $\lambda_j(t) = \lambda_j$), we get

$$R_j(t) = e^{-\lambda_j t}, \tag{3.21}$$

where
$R_j(t)$ is the unit j reliability at time t.

By substituting Equation 3.21 into Equation 3.20, we obtain

$$R_{ss}(t) = e^{-\sum_{j=1}^{n} \lambda_j t}, \tag{3.22}$$

where
$R_{ss}(t)$ is the series system reliability at time t.

By substituting Equation 3.22 into Equation 3.14, we obtain the following equation for the series system mean time to failure:

$$MTTF_{ss} = \int_0^{\infty} e^{-\sum_{j=1}^{n} \lambda_j t} \, dt$$

$$= \frac{1}{\sum_{j=1}^{n} \lambda_j}, \tag{3.23}$$

where

$MTTF_{ss}$ is the series system mean time to failure.

By inserting Equation 3.22 into Equation 3.6, we obtain the following equation for the series system hazard rate:

$$\lambda_{ss}(t) = -\frac{1}{e^{-\sum_{j=1}^{n} \lambda_j t}} \left(-\sum_{j=1}^{n} \lambda_j \right) e^{-\sum_{j=1}^{n} \lambda_j t}$$

(3.24)

$$= \sum_{j=1}^{n} \lambda_j,$$

where

$\lambda_{ss}(t)$ is the series system hazard rate.

The right-hand side of Equation 3.24 is independent of time t. Thus, the left-hand side of this equation is simply λ_{ss}, the failure rate of the series system/configuration/network. It means that whenever we add up failure rates of units/items, we automatically assume that these units/items form a series configuration/network, a worst-case design scenario in regard to reliability.

Example 3.5

Assume that an engineering system is composed of four identical and independent units and that the constant failure rate of each unit is 0.0005 failures per hour. For the successful operation of the engineering system, all the four units must work normally. Calculate the following:

- The engineering system failure rate
- The engineering system reliability for an 11-hour mission
- The engineering system mean time to failure

By inserting the given data values into Equation 3.24, we get

$$\lambda_{ss} = 4(0.0005)$$
$$= 0.002 \text{ failures/hour.}$$

Using the specified data values in Equation 3.22 yields

$$R_{ss}(11) = e^{-(0.0005)(4)(1)}$$
$$= 0.9782.$$

By substituting the given data values into Equation 3.23, we obtain

$$MTTF_{ss} = \frac{1}{4(0.0005)}$$

$$= 500 \text{ hours.}$$

Thus, the engineering system failure rate, reliability, and mean time to failure are 0.002 failures/hour, 0.9782, and 500 hours, respectively.

3.4.2 Parallel Configuration

In the case of parallel configuration, the system is made up of n simultaneously operating units/items, and for the successful operation of the system, at least one of these units/items must operate normally. The n unit parallel system block diagram is shown in Figure 3.3, and each block in the diagram represents a unit.

The parallel system probability of failure, shown in Figure 3.3, is expressed by

$$F_{ps} = P(\bar{E}_1 \bar{E}_2 \bar{E}_3 \ldots \bar{E}_n), \tag{3.25}$$

where
$\bar{E}_j$ is the failure (i.e., failure event) of unit j, for $j = 1, 2, 3, \ldots, n$.
$P(\bar{E}_1 \bar{E}_2 \bar{E}_3 \ldots \bar{E}_n)$ is the occurrence probability of events $\bar{E}_1 \bar{E}_2, \bar{E}_3, \ldots,$ and $\bar{E}_n$.
F_{ps} is the failure probability of the parallel system.

For independently failing parallel units, Equation 3.25 becomes

$$F_{ps} = P(\bar{E}_1) P(\bar{E}_2) P(\bar{E}_3) \ldots P(\bar{E}_n), \tag{3.26}$$

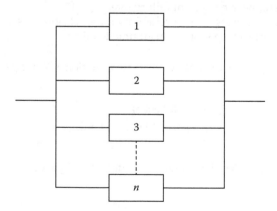

FIGURE 3.3
An n unit parallel system (configuration).

where
$P(\bar{E}_j)$ is the probability of occurrence of failure event $\bar{E}_j$, for $j = 1, 2, 3, ..., n$.

If we let $F_j = P(\bar{E}_j)$ for $j = 1, 2, 3, ..., n$, then Equation 3.26 becomes

$$F_{ps} = F_1 F_2 F_3 ... F_n, \tag{3.27}$$

where
F_j is the failure probability of unit j, for $j = 1, 2, 3, ..., n$.

By subtracting Equation 3.27 from unity, we get

$$R_{ps} = 1 - F_{ps}$$
$$= 1 - F_1 F_2 F_3 ... F_n, \tag{3.28}$$

where
R_{ps} is the parallel system reliability.

For constant failure rate λ_j of unit j, subtracting Equation 3.21 from unity and then inserting it into Equation 3.28, we get

$$R_{ps}(t) = 1 - (1 - e^{-\lambda_1 t})(1 - e^{-\lambda_2 t})(1 - e^{-\lambda_3 t})...(1 - e^{-\lambda_n t}), \tag{3.29}$$

where
$R_{ps}(t)$ is the parallel system reliability at time t.

For identical units, by substituting Equation 3.29 into Equation 3.14, we obtain

$$MTTF_{ps} = \int_0^\infty \left[1 - \left(1 - e^{-\lambda t}\right)^n \right] dt$$
$$= \frac{1}{\lambda} \sum_{j=1}^n \frac{1}{j}, \tag{3.30}$$

where
$MTTF_{ps}$ is the parallel system mean time to failure.
λ is the unit constant failure rate.

Example 3.6

Assume that an engineering system is composed of two independent and identical units in parallel. The constant failure rate of a unit is 0.0008 failures per hour. Calculate the engineering system mean time to failure and reliability for a 15-hour mission.

By inserting the given data values into Equation 3.30, we get

$$MTTF_{ps} = \frac{1}{(0.0008)}\left(1 + \frac{1}{2}\right)$$

$$= 1875 \text{ hours.}$$

By substituting the given data values into Equation 3.29, we obtain

$$R_{ps}(15) = 1 - (1 - e^{-(0.0008)(15)})(1 - e^{-(0.0008)(15)})$$

$$= 0.9998.$$

Thus, engineering system mean time to failure and reliability are 1875 hours and 0.9998, respectively.

3.4.3 *k*-out-of-*n* Configuration

The *k*-out-of-*n* configuration is another type of redundancy in which at least *k* units out of *n* active units must work normally for successful system operation. A *k*-out-of-*n* unit system/configuration block diagram is shown in Figure 3.4. Each block in the diagram denotes a unit. For *k* = 1 and *k* = *n*, the parallel and series configurations are special cases of this configuration, respectively.

For independent and identical units, with the aid of binomial distribution, we write down the following expression for the reliability of *k*-out-of-*n* unit configuration shown in Figure 3.4:

$$R_{k/n} = \sum_{j=k}^{n} \binom{n}{j} R^j (1-R)^{n-j}, \tag{3.31}$$

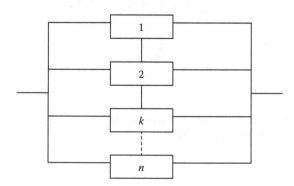

FIGURE 3.4
k-out-of-*n* system/configuration block diagram.

where

$$\binom{n}{j} = \frac{n!}{j!(n-j)!}.$$ (3.32)

$R_{k/n}$ is the k-out-of-n configuration reliability.
R is the unit reliability.

For constant failure rates of identical units, with the aid of Equations 3.31 and 3.11, we obtain

$$R_{k/n}(t) = \sum_{j=k}^{n} \binom{n}{j} e^{-j\lambda t} \left(1 - e^{-\lambda t}\right)^{n-j},$$ (3.33)

where
$R_{k/n}(t)$ is the reliability of k-out-of-n configuration at time t.
λ is the unit constant failure rate.

By inserting Equation 3.33 into Equation 3.14, we get

$$MTTF_{k/n} = \int_0^\infty \left[\sum_{j=k}^{n} \binom{n}{j} e^{-j\lambda t} \left(1 - e^{-\lambda t}\right)^{n-j} \right] dt$$

$$= \frac{1}{\lambda} \sum_{j=k}^{n} \frac{1}{j},$$ (3.34)

where
$MTTF_{k/n}$ is the k-out-of-n configuration mean time to failure.

Example 3.7

An engineering system has four independent and identical units in parallel. At least three units must operate normally for the successful operation of the engineering system. Calculate the engineering system mean time to failure if the unit constant failure rate is 0.0002 failures/hour.
By inserting the given data values into Equation 3.34, we obtain

$$MTTF_{3/4} = \frac{1}{(0.0002)} \left(\frac{1}{3} + \frac{1}{4} \right)$$

$$= 2916.7 \text{ hours.}$$

Thus, the engineering system mean time to failure is 2916.7 hours.

3.4.4 Standby System

The standby system is another reliability configuration/network in which only one unit works and m units are kept in their standby mode. The whole system contains $(m + 1)$ units, and as soon as the working unit malfunctions, the switching mechanism detects the malfunction and turns on one of the standby units. The system fails when all the standby units malfunction. Figure 3.5 shows the block diagram of a standby system with one working and m standby units. Each block in the diagram denotes a unit.

With the aid of Figure 3.5 block diagram, for independent and identical units, time-dependent unit failure rate, and perfect switching mechanism, we write down the following equation for the standby system reliability [6]:

$$R_{ss}(t) = \frac{\sum_{j=0}^{m} \left\{ \left[\int_0^t \lambda(t) dt \right]^j e^{-\int_0^t \lambda(t) dt} \right\}}{j!}, \tag{3.35}$$

where
$R_{ss}(t)$ is the standby system reliability at time t.
$\lambda(t)$ is the unit time-dependent failure rate or hazard rate.

For unit constant failure rate (i.e., $\lambda(t) = \lambda$), Equation 3.35 yields

$$R_{ss}(t) = \frac{\sum_{j=0}^{m} (\lambda t)^j e^{-\lambda t}}{j!}, \tag{3.36}$$

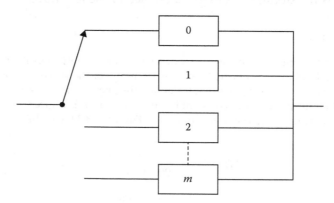

FIGURE 3.5
Standby system block diagram, with one working and m standby units.

where
  λ is the unit constant failure rate.

By inserting Equation 3.36 into Equation 3.14, we get

$$MTTF_{ss} = \int_0^{\infty} \left[\frac{\sum_{j=0}^{m} (\lambda t)^j e^{-\lambda t}}{j!} \right] dt \qquad (3.37)$$

$$= \frac{m+1}{\lambda},$$

where
  $MTTF_{ss}$ is the standby system mean time to failure.

Example 3.8

Assume that an engineering standby system contains two independent and identical units (i.e., one working; the other on standby). The unit constant failure rate is 0.002 failures/hour.

Calculate the standby system reliability for a 60-hour mission and mean time to failure, assuming that the standby unit remains as good as new in its standby mode and the switching mechanism is perfect.

By substituting the specified data values into Equation 3.36, we get

$$R_{ss}(60) = \frac{\sum_{j=0}^{1} \left\{ \left[(0.002)(60) \right]^j e^{-(0.002)(60)} \right\}}{j!}$$

$$= 0.9933.$$

Similarly, by inserting the given data values into Equation 3.37, we obtain

$$MTTF_{ss} = \frac{2}{(0.002)}$$

$$= 1000 \text{ hours.}$$

Thus, the standby system reliability and mean time to failure are 0.9933 and 1000 hours, respectively.

3.4.5 Bridge Configuration

In some engineering systems, units may form a bridge configuration, as shown in Figure 3.6. Each block in Figure 3.6 denotes a unit, and all units are labeled with numerals.

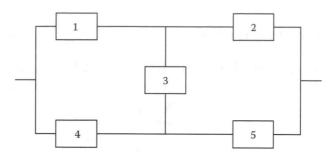

FIGURE 3.6
A five dissimilar unit bridge configuration.

For independent units, Figure 3.6 reliability is expressed by Lipp [7] as

$$
\begin{aligned}
R_{bc} = {} & 2R_1R_2R_3R_4R_5 + R_1R_3R_5 + R_2R_3R_4 \\
& + R_2R_5 + R_1R_4 - R_1R_2R_3R_4 - R_1R_2R_3R_5 \\
& - R_2R_3R_4R_5 - R_1R_2R_4R_5 - R_3R_4R_5R_1,
\end{aligned} \tag{3.38}
$$

where
R_j is the *j*th unit reliability, for $j = 1, 2, 3, 4,$ and 5.
R_{bc} is the bridge configuration reliability.

For identical units, Equation 3.38 becomes

$$
R_{bc} = 2R^5 - 5R^4 + 2R^3 + 2R^2, \tag{3.39}
$$

where
R is the unit reliability.

For constant unit failure rate, with the aid of Equations 3.11 and 3.39, we obtain

$$
R_{bc}(t) = 2e^{-5\lambda t} - 5e^{-4\lambda t} + 2e^{-3\lambda t} + 2e^{-2\lambda t}, \tag{3.40}
$$

where
$R_{bc}(t)$ is the bridge configuration reliability at time *t*.
λ is the unit constant failure rate.

By substituting Equation 3.40 into Equation 3.14, we obtain

$$
MTTF_{bc} = \frac{49}{60\lambda}, \tag{3.41}
$$

where

$MTTF_{bc}$ is the bridge configuration/network mean time to failure.

Example 3.9

Assume that an engineering system has five identical and independent units forming a bridge configuration/network. The constant failure rate of each unit is 0.0005 failures/hour. Calculate the bridge configuration mean time to failure and reliability for a 400-hour mission.

By inserting the given data value into Equation 3.41, we get

$$MTTF_{bc} = \frac{49}{60(0.0005)}$$

$$= 1633.3 \text{ hours.}$$

Similarly, by inserting the specified data values into Equation 3.40, we obtain

$$R_{bc} = 2e^{-5(0.0005)(400)} - 5e^{-4(0.0005)(400)} + 2e^{-3(0.0005)(400)} + 2e^{-2(0.0005)(400)} = 0.9273.$$

Thus, the bridge configuration mean time to failure and reliability are 1633.3 hours and 0.9273, respectively.

3.5 The Need for Safety and the Role of Engineers in Regard to Safety

The desire to be safe and secure has always been an important concern to humans. For example, early humans took appropriate precautions for guarding against natural hazards around them. Moreover, in 2000 BC, an ancient Babylonian ruler named Hammurabi developed a code known as the *Code of Hammurabi*. The code included clauses on items such as allowable fees for physicians and monetary damages against individuals who caused injury to others [8,9].

Nowadays, safety has become a very important issue because every year, a large number of people die and get seriously injured due to workplace-related and other accidents. For example, as per the National Safety Council, in 1996, in the United States, there were 93,400 deaths and a very large number of disabling injuries due to accidents [10]. The cost of these accidents was estimated to be approximately US$121 billion.

Some of the factors that play a pivotal role in demanding better safety are public pressure, government regulations, and an increasing number of lawsuits.

Nowadays, engineering systems/products have become very sophisticated and complex. The matter of safety concerning such systems/products has become a very challenging issue for engineers who are constantly pressured to complete new designs rapidly and at lower costs due to competition and other factors. Experiences over the years indicate that this, in turn, usually leads to more design-related deficiencies, errors, and causes of accidents.

Design-related deficiencies or shortcomings can cause or contribute to accidents. These deficiencies or shortcomings may result because a designer/design [11]

- Fails to foresee an expected application of a system/product/item or its potential consequences
- Violates general capabilities/tendencies of potential users
- Fails to eradicate or reduce the occurrence of human error
- Creates an arrangement of operating controls and other devices that quite significantly increases reaction time during an emergency or is conducive to the occurrence of errors
- Places an unreasonable level of stress on potential users/operators
- Offers incorrect, confusing, or unfinished concepts/products
- Fails to appropriately warn of a potential hazard
- Creates an unsafe characteristic of a system/product/item
- Fails to provide an acceptable level of protection in a user's personal protective equipment
- Does not appropriately consider or determine the consequences of an error, an omission, an action, or a failure
- Fails to prescribe a proper operational procedure in situations where a hazard might exist
- Relies on system/product/item users for avoiding an accident
- Incorporates weak warning mechanisms rather than providing a safe design to eliminate hazards.

3.6 Product Hazard Classifications

There are many product-related hazards. They may be grouped under six classifications as shown in Figure 3.7 [12].

Classification I: Electrical hazards have two main elements: electrocution hazard and shock hazard. The major electrical hazard to system/product stems from electrical faults, frequently referred to as *short circuits*. Classification II: Energy hazards may be divided into two categories: kinetic

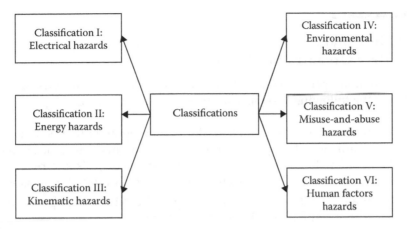

FIGURE 3.7
Product hazard classifications.

energy hazards and potential energy hazards. The kinetic energy-related hazards pertain to items that have energy due to their motion. Two examples of these items are flywheels and fan blades. Any object that interferes with such items' motion can experience extensive damage. The potential energy-related hazards pertain to items that store energy. Three examples of these items are electronic capacitors, springs, and counterbalancing weights. During equipment servicing, such hazards are very important because stored energy, when released, can suddenly result in serious injury.

Classification III: Kinematic hazards pertain to situations where parts/ items come together while moving and result in pinching, cutting, or crushing an object/item caught between them. Classification IV: Environmental hazards may be divided into two categories: external hazards and internal hazards. External hazards are the hazards posed by the system product during its life span and include items such as disposal hazards, maintenance-related hazards, and service-life operation hazards. The internal hazards are concerned with the changes in the surrounding environment that result in an internally damaged product/system/item. A careful consideration to factors such as extremes of temperatures, vibrations, electromagnetic radiation, atmospheric contaminants, and ambient noise level during the design phase can be very helpful for eliminating or minimizing the internal hazards.

Classification V: Misuse-and-abuse hazards are concerned with the usage of a product/system by humans. Misuse of a product/system can cause serious injuries, and its abuse can lead to injuries or hazardous situations. Two examples of the causes for product/system abuse are poor operating practices and lack of proper maintenance. Classification VI: Human factors hazards are concerned with poor design in regard to humans, that is, to their length of reach, physical strength, weight, height, visual angle, visual acuity, intelligence, and computational ability, etc.

3.7 Safety Management Principles and Product Safety Organization Tasks

There are many principles of safety management. The main ones are presented in the following [13–15]:

- Safety should be managed just like managing any other activity in an organization/company. More specifically, management should direct safety by setting attainable safety-related goals and by planning, organizing, and controlling to successfully attain the set goals.
- The main activity of safety is finding and defining the operational errors that result in accidents.
- The safety system should be tailored to effectively fit the organization/company culture.
- The causes leading to unsafe behavior can be identified, classified, and controlled.
- Under most circumstances, unsafe behavior is a normal behavior because it is the result of normal human beings reacting to the environment surrounding them. Therefore, it is clearly the management's responsibility to make appropriate changes to the environment that leads to the unsafe behavior.
- In developing a good safety system, the main subsystems that must be considered are the managerial, the physical, and the behavioral.
- The important symptoms that highlight that something is not right in the management system are an unsafe act, an unsafe condition, and an accident.
- There is no single approach for achieving safety in an organization/company. But for a safety system to be effective, it must clearly satisfy certain criteria: have the top-level management visibly showing its support, be flexible, and involve workers' participation, etc.
- There are certain sets of conditions that can be predicted to lead to severe injuries: high energy sources, unusual, nonroutine tasks, certain construction conditions, and nonproductive activities.
- The key to effective line safety performance is management procedures that clearly and effectively factor in accountability.

A product safety organization performs a variety of tasks. The main ones are as follows [15–17]:

- Review warning labels that are to be placed on the system/product in regard to satisfying all legal requirements, compatibility to warnings in the instruction manuals, and adequacy.

- Review all safety-related customer complaints and field reports.
- Review all governmental and nongovernmental system/product safety-related requirements.
- Develop a system by which the safety program can be monitored effectively.
- Prepare the system/product safety-related directives and program.
- Develop safety criteria on the basis of all applicable governmental and voluntary standards for use by organization/company, vendor, and subcontractor design professionals.
- Provide assistance to designers in selecting alternative means for controlling or eradicating hazards or other safety-associated problems in preliminary designs.
- Review system/product test reports for determining shortcomings or trends with respect to safety.
- Review proposed system/product operation and maintenance-related documents in regard to safety.
- Determine if items, such as protective equipment, warning and monitoring devices, or emergency equipment, are really needed for the system/product.
- Participate in reviewing accident-related claims or recall actions by government agencies/bodies and recommend appropriate remedial measures for justifiable recalls or claims.
- Review all types of hazards and mishaps in current similar system/ product for avoiding their repetition in the new systems/products.
- Review the system/product for determining if all potential hazards have been appropriately controlled or eradicated.

3.8 Accident Causation Theories

There are many accident causation theories [9]. Two of these theories are described in the following.

3.8.1 Human Factors Accident Causation Theory

The basis for the human factors accident causation theory is the assumption that accidents occur due to a chain of events directly or indirectly due to human error. The theory consists of three main factors shown in Figure 3.8 that lead to the occurrence of human error [9,18].

The factor *overload* is concerned with the imbalance between a person's capacity at any point in time and the amount of load he or she is carrying in a

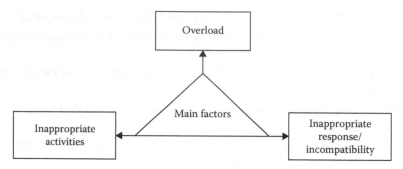

FIGURE 3.8
Main factors leading to the occurrence of human error.

given state. The capacity of a person is the product of many factors including natural ability, stress, state of mind, degree of training, physical condition, and fatigue. The load carried by a person is composed of tasks for which he or she has responsibility along with additional burdens resulting from the situational factors (i.e., level of risk, unclear instructions, etc.), internal factors (i.e., personal problems, worry, etc.), and environmental factors (i.e., distractions, noise, etc.).

The factor *inappropriate activities* is concerned with inappropriate activities carried out by a person due to human error. For example, a person misjudged the degree of risk involved in a stated task and then conducted the task on that misjudgment.

The factor *inappropriate response/incompatibility* is another major human error causal factor, and three examples of inappropriate response by a person are as follows [15,18]:

- A person disregarded the stated safety procedures.
- A person removed a safeguard from equipment/machine for improving output.
- A person detected a hazardous condition but took no necessary corrective action.

Additional information on this theory is available in the study by Heinrich et al. [18].

3.8.2 Domino Accident Causation Theory

The domino accident causation theory is encapsulated in 10 statements by H. W. Heinrich, called the *axioms of industrial safety*, presented in the following [19]:

- *Statement 1:* Supervisors play a key role in industrial accident prevention.
- *Statement 2:* An unsafe condition or an unsafe act by a person does not always immediately lead to an accident/injury.

- *Statement 3:* Most accidents are the result of unsafe acts of people.
- *Statement 4:* An accident can occur only when a person commits an unsafe act and/or there is a physical- or mechanical-related hazard.
- *Statement 5:* Management should assume full safety responsibility with vigor because it is in the best position for achieving results effectively.
- *Statement 6:* There are two types of costs of an accident: direct and indirect. Three examples of the direct cost are liability claims, compensation, and medical costs.
- *Statement 7:* The occurrence of injuries results from a completed sequence of a number of factors; the last or the final one of which is the accident itself.
- *Statement 8:* The reasons why humans commit unsafe acts can be useful in selecting appropriate corrective measures.
- *Statement 9:* The severity of an injury is largely fortuitous, and the specific accident that caused it is generally preventable.
- *Statement 10:* The most helpful accident-prevention methods are analogous to the productivity and quality approaches.

Heinrich believed that there are five factors, presented in Table 3.1, in the sequence of events leading up to an accident [9,17].

In factor Ancestry and Social Environment, it is assumed that negative character traits such as recklessness, stubbornness, and avariciousness that might lead people to behave in an unsafe manner can be inherited through one's ancestry or acquired as a result of the social environment or surroundings. In factor *fault of person*, it is assumed that negative character traits (whether inherited or acquired) such as recklessness, violent temper, nervousness, excitability, and ignorance of safety-related practices constitute proximate reasons for committing unsafe acts or for the existence of physical or mechanical hazards.

In factor *unsafe act/mechanical or physical hazard*, it is assumed that unsafe acts by humans (starting equipment/machinery without warning, removing safeguards, standing under suspended loads) and mechanical or physical hazards (unguarded gears, inadequate light, absence of guard rails, unguarded point of

TABLE 3.1

Factors in the Sequence of Events Leading up to an Accident

Factor No.	Factor Description
1	Ancestry and social environment
2	Fault of person
3	Unsafe act/mechanical or physical hazard
4	Accident
5	Injury

operation) are the direct causes for the accidents' occurrence. In factor *accident*, it is assumed that events such as falls of humans and striking of humans by flying objects are the typical examples of accidents that lead to injury.

Finally, in factor *injury*, it is assumed that typical injuries directly resulting from the accidents' occurrence include fractures and lacerations.

3.9 Facts and Figures Related to Engineering Maintenance

Some of the facts and the figures directly or indirectly related to engineering maintenance are as follows:

- As per 1997 Department of Defense (DOD) Budget [20], for the fiscal year 1997, the request of the US Department of Defense (DOD) for its operation and maintenance budget was US$79 billion.
- The US industrial sector spends over US$300 billion annually on plant operations and maintenance [21].
- As per *Report by the Working Party on Maintenance Engineering* [22] and Kelly [23], in 1970, British Ministry of Technology working committee document reported that the UK annual maintenance cost was approximately £3000 million.
- The annual cost of maintaining a military jet aircraft is about US$1.6 million, and approximately 11% of the operating cost for an aircraft accounts for maintenance-related activities [24].
- As per *Report on Infrastructure and Logistics* [25], the US DOD spends about US$12 billion per year on depot maintenance of weapon systems and equipment.
- As per Niebel [26], over the years, the size of a plant maintenance group in a manufacturing organization has varied from 5% to 10% of the entire operating force.

3.10 Maintenance Engineering Objectives

There are many maintenance engineering objectives. Eight of the important ones are as follows [27,28]:

- *Objective I:* Reduce the frequency and the amount of maintenance.
- *Objective II:* Improve the maintenance-related operations.
- *Objective III:* Reduce the amount of supply support required.

- *Objective IV:* Decrease the maintenance skills required.
- *Objective V:* Improve the maintenance organization.
- *Objective VI:* Improve and ensure maximum usage of all maintenance facilities.
- *Objective VII:* Establish optimum frequency and extent of preventive maintenance to be carried out.
- *Objective VIII:* Reduce the effect of complexity.

3.11 Preventive Maintenance

Preventive maintenance is an important component of a maintenance activity, and within a maintenance department, it generally accounts for a significant proportion of the overall maintenance activity. It is the care and the servicing by maintenance personnel for keeping facilities in a satisfactory operational state by providing for systematic detection, inspection, and correction of incipient failures either prior to their development into major failures or prior to their occurrence [27,28].

This section presents various important aspects of preventive maintenance.

3.11.1 Preventive Maintenance Elements and Principle for Selecting Items for Preventive Maintenance

There are seven elements for preventive maintenance [27,28]:

- *Element I: Inspection*—Element I is concerned with periodically inspecting items/units for determining their serviceability by comparing their mechanical, electrical, physical, and other characteristics to established standards.
- *Element II: Adjustment*—Element II is concerned with periodically making adjustments to stated variable elements for achieving optimum performance.
- *Element III: Alignment*—Element III is concerned with making changes to an item's stated variable elements for achieving optimum performance.
- *Element IV: Calibration*—Element IV is concerned with detecting and adjusting any discrepancy in the accuracy of the parameter or the material being compared to the established standard value.
- *Element V: Servicing*—Element V is concerned with periodically charging, cleaning, lubricating, and so on items/materials for preventing the occurrence of incipient failures.

- *Element VI: Testing*—Element VI is concerned with periodically testing for determining serviceability and detecting electrical or mechanical degradation.
- *Element VII: Installation*—Element VII is concerned with periodically replacing limited-life items or items experiencing time cycle or wear degradation, for maintaining the stated tolerance level.

The formula principle presented in the following can be very helpful in deciding whether to implement a preventive maintenance program for a system/item [29,30].

$$(m)(AC)(\alpha) > SPMC, \tag{3.42}$$

where
  $SPMC$ is the total cost of the system preventive maintenance.
  m is the total number of breakdowns.
  AC is the average cost per breakdown.
  α is 70% of the total cost of breakdowns.

3.11.2 Steps for Developing Preventive Maintenance Program

The development of a good preventive maintenance program requires the availability of a number of items including accurate historical records of equipment, past data from similar equipment, test instruments and tools, manufacturer's recommendations, skilled personnel, management support and user cooperation, and service manuals [31]. A good preventive maintenance program can be developed in a short time by following the six steps presented in the following [32]:

- *Step 1: Highlight and choose the areas*—Step 1 is concerned with highlighting and choosing one or two important areas on which to concentrate the initial preventive effort. The main objective of this step is to obtain good results in highly visible areas.
- *Step 2: Identify the preventive maintenance requirements*—Step 2 is concerned with defining the preventive maintenance-related needs and then developing a schedule for two types of tasks: periodic preventive maintenance assignments and daily preventive maintenance inspections.
- *Step 3: Determine assignment frequency*—Step 3 is concerned with establishing the frequency of assignments and reviewing the item/equipment conditions and records. The frequency depends on various factors including the experience of personnel familiar with the

equipment/item under consideration, recommendations from engineers, and vendor recommendations.

- *Step 4: Prepare the preventive maintenance assignments*—Step 4 is concerned with preparing the periodic and daily assignments effectively and then getting them approved.

- *Step 5: Schedule the preventive maintenance assignments*—Step 5 is concerned with scheduling the defined preventive maintenance assignments on the basis of a 12-month period.

- *Step 6: Expand the preventive maintenance program as appropriate*—Step 6 is concerned with expanding the preventive maintenance program to other areas on the basis of experience/knowledge gained from the pilot preventive maintenance projects.

3.11.3 Preventive Maintenance Measures

There are many preventive maintenance-related measures. Two such measures considered quite useful are presented in the following [27,28,33]:

- *Mean preventive maintenance time*

  Mean preventive maintenance time is the average system/equipment downtime required for performing scheduled preventive maintenance. Mean preventive time is defined by

$$MPMT = \frac{\sum_{i=1}^{m} APMT_i f_i}{\sum_{i=1}^{m} f_i}, \tag{3.43}$$

where
  $MPMT$ is the mean preventive maintenance time.
  m is the total number of data points.
  f_i is the frequency of i preventive maintenance task in tasks per operating hour after adjustment for equipment/item duty cycle.
  $APMT_i$ is the average time required to perform preventive maintenance task i, for $i = 1, 2, 3, \ldots, m$.

- *Median preventive maintenance time*

  Median preventive maintenance time is the equipment/item downtime required for performing 50% of all scheduled preventive maintenance actions under the conditions specified for median preventive maintenance time.

For lognormal distributed preventive maintenance times, the median preventive maintenance time is expressed by

$$MPMT = \text{anti} \log \left(\frac{\sum_{i=1}^{m} \lambda_i \log APMT_i}{\sum_{i=1}^{m} \lambda_i} \right), \tag{3.44}$$

where

$MPMT$ is the median preventive maintenance time.

m is the total number of data points.

λ_i is the constant failure rate of component i of the equipment/item for which maintainability is to be determined, adjusted for factors such as tolerance and interaction failures, duty cycle, and catastrophic failures that will lead to the deterioration of equipment/item performance to the degree that a maintenance-related action will be taken for $i = 1, 2, 3, ..., m$.

$APMT_i$ is the average time required to perform preventive maintenance task i, for $i = 1, 2, 3, ..., m$.

3.11.4 Preventive Maintenance Benefits and Drawbacks

There are many benefits of performing preventive maintenance. Most of the important ones are as follows [29,31,34].

- Improved safety
- Increment in equipment/system availability
- Reduction in need for standby equipment/system
- Reduction in parts inventory
- Stimulation in preaction instead of reaction
- Consistency in quality
- Increment in production revenue
- Performed as convenient
- Standardized procedures, costs, and times
- Reduction in overtime
- Useful in promoting cost/benefit optimization
- Balanced workload

In contrast, some of the drawbacks of performing preventive maintenance are exposing equipment/system to possible damage, use of more parts/components, more frequent access to equipment/system, and increase in initial costs [29,31,34].

3.12 Corrective Maintenance

Corrective maintenance is an important element of overall maintenance activity and is the remedial action performed because of failure or deficiencies discovered during preventive maintenance or, otherwise, repair an item/equipment to its operating state [27,28,34–36]. Usually, corrective maintenance is an unplanned maintenance activity that needs immediate attention that must be added, integrated with, or substituted for earlier scheduled work.

This section presents various important aspects of corrective maintenance.

3.12.1 Types of Corrective Maintenance

Corrective maintenance may be grouped under five classifications shown in Figure 3.9 [27,34,37].

Classification I: Fail repair is concerned with restoring the equipment/item to its operational state. Classification II: Servicing may be needed because of a corrective maintenance action (e.g., engine repair can result in need for crankcase refill, welding on). Classification III: Overhaul is concerned with restoring or repairing equipment/item to its complete serviceable state

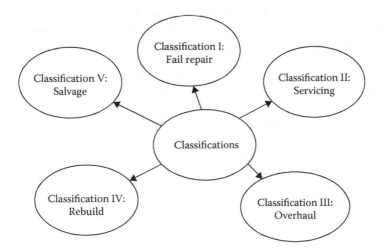

FIGURE 3.9
Corrective maintenance classifications.

meeting requirements stated in maintenance serviceability standards, using the inspect and repair only as appropriate approach.

Classification IV: Rebuild is concerned with restoring equipment/item to a standard as close as possible to its original state in regard to appearance, life expectancy, and performance. This is fulfilled through actions such as examination of all parts/components, complete disassembly, repair or replacement of unserviceable or worn components as per original specifications and manufacturing-related practices, and reassembly and testing to original production-related requirements. Classification V: Salvage is concerned with the disposal of nonrepairable materials and utilization of salvaged materials from items that are impossible to repair in the overhaul, rebuild, or repair programs.

3.12.2 Corrective Maintenance Steps, Downtime Components, and Time Reduction Strategies at System Level

Over the years, different researchers and authors have proposed different steps for performing corrective maintenance. Nonetheless, corrective maintenance can be performed in the five steps presented in the following [27]:

- *Step 1: Failure recognition*—Step 1 is concerned with recognizing the existence of a failure.
- *Step 2: Failure location*—Step 2 is concerned with localizing the failure within the system to a specific piece of equipment/item.
- *Step 3: Diagnosis within the equipment/item*—Step 3 is concerned with diagnosis within equipment/item for identifying specific failed part or component.
- *Step 4: Failed part replacement or repair*—Step 4 is concerned with replacing or repairing failed parts/components.
- *Step 5: Return system/equipment to service*—Step 5 is concerned with checking out and returning the system/equipment back to service.

Corrective maintenance downtime is made up of three major components: active repair time, delay time, and administrative and logistic times [27,38]. The six subcomponents of the active repair time are fault correction time, checkout time, adjustment and calibration times, fault location time, spare item obtainment time, and preparation time.

In order to improve the effectiveness of corrective maintenance, it is very important to reduce corrective maintenance time. The following five strategies are considered quite useful to reduce system-level corrective maintenance [28,33,34]:

- *Improve fault recognition, isolation, and location:* Experiences over the years clearly indicate that within a corrective maintenance activity, fault recognition, isolation, and location consume the most time. The factors that are helpful to reduce corrective maintenance time are well-designed fault indicators, good maintenance procedures, unambiguous fault isolation capability, and well-trained maintenance personnel.

- *Improve accessibility:* Experiences over the years clearly indicate that, frequently, a significant amount of time is spent accessing failed parts/components. Careful attention to accessibility during the design process can help reduce the accessibility time of parts/components and, consequently, the corrective maintenance-related time.

- *Employ redundancy:* Employing redundancy is concerned with designing in appropriate redundant parts/components that can be switched on during the repair process of faulty parts/components so that the equipment/system continues to function. In this situation, although the overall maintenance-related workload may not be reduced, the equipment/system downtime could be impacted quite significantly.

- *Improve interchangeability:* Proper physical and functional interchangeability is a very important factor in removing and replacing parts/components, thus reducing corrective maintenance time.

- *Consider human factors:* During the design process, paying proper attention to human factors such as size, shape, and weight of components/parts; readability of instructions; selection and placement of dials and indicators; size and placement of access and gates; and information-processing aids can help lower corrective maintenance time quite significantly.

3.12.3 Corrective Maintenance Measures

There are many corrective maintenance measures. Two such measures considered quite useful are as follows [27,33,34,39]:

- *Mean corrective maintenance time*

  Mean corrective maintenance time is an important corrective maintenance measure and is expressed by

$$MCMT = \frac{\sum_{j=1}^{m} \lambda_j CMT_j}{\sum_{j=1}^{m} \lambda_j},$$
(3.45)

where

$MCMT$ is the mean corrective maintenance time.

m is the total number of equipment parts/elements.

CMT_j is the equipment jth part/element corrective maintenance time, for $j = 1, 2, 3, ..., m$.

λ_j is the equipment jth part/element constant failure rate, for $j = 1, 2, 3, ..., m$.

Normally, corrective maintenance times are described by exponential, normal, and lognormal probability distributions. Examples of the types of equipment that follow these probability distributions are presented in the following:

- *Exponential distribution:* Often, corrective maintenance times of electronic equipment with a good built-in test capability and rapid remove-and-replace maintenance concept follow exponential distribution.

- *Normal distribution:* Often, corrective maintenance times of electromechanical or mechanical equipment with a remove-and-replacement maintenance concept follow normal distribution.

- *Lognormal distribution:* Usually, corrective maintenance times of electronic equipment that does not possess built-in test capability follow lognormal distribution.

- *Median active correction maintenance time*

    Median active correction maintenance time is another important corrective maintenance measure, and it normally provides the best average location of the sample data and is the 50th percentile of all corrective maintenance time values. Median active corrective maintenance time is a measure of the time within which 50% of all corrective maintenance-related activities can be carried out. The computation of this specific measure is subject to the probability distribution describing corrective maintenance times.

    Thus, the median of corrective maintenance times following a lognormal probability distribution is defined by

$$ACMT_\mathrm{m} = \text{anti} \log \left(\frac{\sum\limits_{j=1}^{m} \lambda_j \log CMT_j}{\sum\limits_{j=1}^{m} \lambda_j} \right), \tag{3.46}$$

where $ACMT_\mathrm{m}$ is the active corrective maintenance times median.

PROBLEMS

1. Describe the bathtub hazard rate curve and write down the equation that can be used to represent it.
2. Write down general equations for the following:
   a. Failure density function
   b. Reliability function
   c. Hazard rate
3. Write down three different formulas that can be used to obtain mean time to failure expressions for engineering systems.
4. Assume that an engineering system is composed of five independent and identical units and the constant failure rate of each unit is 0.0004 failures per hour. For the successful operation of the engineering system, all the five units must operate normally. Calculate the following:
   a. The engineering system mean time to failure
   b. The engineering system failure rate
   c. The engineering system reliability for a 15-hour mission
5. What are the special case configurations of the k-out-of-n configuration? Write down their mean time to failure expressions for identical units.
6. What are the design-related deficiencies in engineering systems that can cause or contribute to accidents?
7. What are the safety management principles?
8. Describe human factors accident causation theory.
9. Discuss at least eight important objectives of maintenance engineering.
10. What are the advantages and the disadvantages of preventive maintenance?

References

1. Kapur, K.C., Reliability and maintainability, in *Handbook of Industrial Engineering*, edited by G. Salvendy, Wiley, New York, 1982, pp. 8.5.1–8.5.34.
2. Dhillon, B.S., *Design Reliability: Fundamentals and Applications*, CRC Press, Boca Raton, FL, 1999.
3. Dhillon, B.S., Life distributions, *IEEE Transactions on Reliability*, Vol. 30, No. 5, 1981, pp. 457–460.
4. Shooman, M.L., *Probabilistic Reliability: An Engineering Approach*, McGraw-Hill, New York, 1968.

5. Dhillon, B.S., *Reliability, Quality, and Safety for Engineers*, CRC Press, Boca Raton, FL, 2005.
6. Sandler, G.H., *System Reliability Engineering*, Prentice Hall, Englewood Cliffs, NJ, 1963.
7. Lipp, J.P., Topology of switching elements versus reliability, *Transactions on IRE Reliability and Quality Control*, Vol. 7, 1957, pp. 21–34.
8. Ladon, J., Editor, *Introduction to Occupational Health and Safety*, National Safety Council (NSC), Chicago, 1986.
9. Goetsch, D.L., *Occupational Safety and Health*, Prentice Hall, Englewood Cliffs, NJ, 1996.
10. *Accidental Facts*, Report, National Safety Council, Chicago, 1996.
11. Hammer, W., Price, D., *Occupational Safety Management and Engineering*, Prentice Hall, Upper Saddle River, NJ, 2001.
12. Hunter, T.A., *Engineering Design for Safety*, McGraw-Hill, New York, 1992.
13. Petersen, D., *Techniques of Safety Management*, McGraw-Hill Book Company, New York, 1971.
14. Petersen, D., *Safety Management*, American Society of Safety Engineers, Des Plaines, IL, 1998.
15. Dhillon, B.S., *Safety and Human Error in Engineering Systems*, CRC Press, Boca Raton, FL, 2013.
16. Hammer, W., *Product Safety Management and Engineering*, Prentice Hall, Englewood Cliffs, NJ, 1980.
17. Dhillon, B.S., *Engineering Safety: Fundamentals, Techniques, and Applications*, World Scientific Publishing, River Edge, NJ, 2003.
18. Heinrich, H.W., Petersen, D., Roos, N., *Industrial Accident Prevention*, McGraw-Hill, New York, 1980.
19. Heinrich, H.W., *Industrial Accident Prevention*, McGraw-Hill, New York, 1959.
20. US General Accounting Office, 1997 DOD budget: Potential reductions to operation and maintenance program, US General Accounting Office, Washington, DC, 1996.
21. Latino, C.J., *Hidden Treasure: Eliminating Chronic Failures Can Cut Maintenance Costs up to 60%*, Reliability Center, Hopewell, VA, 1999.
22. *Report by the Working Party on Maintenance Engineering*, Department of Industry, London, 1970.
23. Kelly, A., *Management of Industrial Maintenance*, Newes-Butterworths, London, 1978.
24. Kumar, V.D., New trends in aircraft reliability and maintenance measures, *Journal of Quality in Maintenance Engineering*, Vol. 5, No. 4, 1999, pp. 287–299.
25. *Report on Infrastructure and Logistics*, US Department of Defense, Washington, DC, 1995.
26. Niebel, B.W., *Engineering Maintenance Management*, Marcel Dekker, New York, 1994.
27. AMCP 706-132, *Engineering Design Handbook: Maintenance Engineering Techniques*, Department of the Army, Washington, DC, 1975.
28. Dhillon, B.S., *Engineering Maintenance: A Modern Approach*, CRC Press, Boca Raton, FL, 2002.
29. Levitt, J., Managing preventive maintenance, *Maintenance Technology*, February 1997, pp. 20–30.
30. Levitt, J., *Maintenance Management*, Industrial Press, New York, 1997.

31. Patton, J.D., *Preventive Maintenance*, Instrument Society of America, Research Triangle Park, NC, 1983.
32. Westerkemp, T.A., *Maintenance Manager's Standard Manual*, Prentice Hall, Paramaus, NJ, 1997.
33. Blanchard, B.S., Verma, D., Peterson, E.L., *Maintainability*, Wiley, New York, 1995.
34. Dhillon, B.S., *Maintainability, Maintenance, and Reliability for Engineers*, CRC Press, Boca Raton, FL, 2006.
35. McKenna, T., Oliverson, R., *Glossary of Reliability and Maintenance Terms*, Gulf Publishing, Houston, TX, 1997.
36. Omdahl, R.P., *Reliability, Availability, and Maintainability (RAM) Dictionary*, ASQC Quality Press, Milwaukee, Wisconsin, 1988.
37. MICOM 750-8, *Maintenance of Supplies and Equipment*, Department of Defense, Washington, DC, March 1972.
38. NAVORD OD 39223, *Maintainability Engineering Handbook*, Department of Defense, Washington, DC, June 1969.
39. AMCP-766-133, *Engineering Design Handbook: Maintainability Engineering Theory and Practice*, Department of Defense, Washington, DC, 1976.

4

Methods for Performing Reliability, Safety, and Maintenance Analysis of Engineering Systems

4.1 Introduction

Over the years, a large amount of published literature in the areas of reliability, safety, and maintenance has appeared in the form of books, technical reports, journal articles, and conference proceeding articles [1–7]. Many of these publications report the development of various types of methods and approaches for performing reliability, safety, and maintenance analyses. Some of these methods and approaches can be used quite effectively for performing analysis in all these three areas. The others are more confined to a specific area (i.e., reliability, safety, or maintenance).

Two examples of the methods and approaches that can be used to perform analysis in reliability, safety, and maintenance areas are the Markov method and the fault tree analysis (FTA). The Markov method is named after a Russian mathematician, Andrei A. Markov (1856–1922), and is a highly mathematical approach that is frequently used for performing various types of reliability, safety, and maintenance analyses of engineering systems. The FTA method was developed in the early 1960s for analyzing the safety of rocket launch control systems in the United States. Today, both the Markov method and FTA are being used across many diverse areas for analyzing various types of problems.

This chapter presents a number of methods considered useful for performing reliability, safety, and maintenance analysis of engineering systems.

4.2 Fault Tree Analysis

Fault tree analysis is a method widely used in industry for evaluating the reliability of engineering systems during their design and development phase, particularly in the area of nuclear power generation. A fault tree may be described as a logical representation of the relationship of fundamental/basic fault events that lead to a stated undesirable event, called the *top event*, and is depicted using a tree structure with logic gates such as OR and AND gates.

The FTA method was developed in the early 1960s at the Bell Telephone Laboratories for performing the analysis of the Minuteman Launch Control System [1]. The main objectives of performing FTA are as follows [1,5].

- To understand the functional relationships of system failures
- To verify the ability of the system to meet its imposed safety-associated requirements
- To highlight cost-effective improvements and critical areas
- To comprehend the degree of protection that the design concept provides against the occurrence of failures
- To meet jurisdictional requirements

Six of the main prerequisites associated with FTA are presented in Table 4.1 [1].

FTA begins by highlighting an undesirable event, called the top event, associated with an item/system under consideration. Fault events that can cause the occurrence of a top event are generated and connected by logic operators such as OR and AND. The OR gate provides a true output (i.e., fault) when one or more of its inputs are true. Similarly, the AND gate provides a true output (i.e., fault) when all its inputs are true.

The construction of a fault tree proceeds by generating fault events in a successive manner until the fault events need not be developed any further. These fault events are known as *primary or basic events*. A fault tree is a logic structure that relates the top fault event to the basic/primary fault events. During the construction of a fault tree, one question that is successively raised is, How could this fault event occur?

TABLE 4.1

Six of the Main Prerequisites Associated with FTA

No.	Prerequisite
1	Clearly defined analysis objectives and scope
2	Clear identification of all associated assumptions
3	Clear understanding of design, operation, and maintenance aspects of item/system under consideration
4	A comprehensive review of item/system operational experience
5	Clearly defined item/system interfaces and item/system physical bounds
6	Clear definition of what constitutes item/system failure: the undesirable event

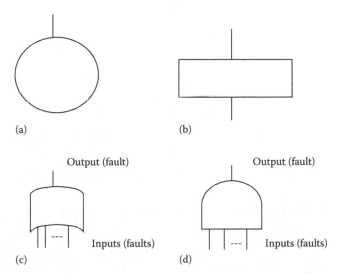

FIGURE 4.1
Basic fault tree symbols: (a) basic fault event, (b) resultant fault event, (c) OR gate, and (d) AND gate.

Figure 4.1 shows four basic symbols used for constructing fault trees. The meanings of gates/symbols OR and AND, shown in Figure 4.1, have already been discussed. The remaining two symbols (i.e., rectangle and circle) are described in the following:

- Rectangle: The rectangle represents a resultant fault event that occurs from the combination of fault events through the input of a logic gate such as AND and OR.
- Circle: The circle represents a primary or a basic fault event (e.g., failure of an elementary component/part), and the basic fault event parameters are failure probability, failure rate, repair rate, and unavailability.

Example 4.1

Assume that a windowless room contains one switch and two light bulbs. Develop a fault tree for the top (undesired) fault event, *dark room*, if the switch can only fail to close.

In this case, there can only be no light in the room (i.e., dark room) if both the light bulbs burn out, if there is no incoming electricity, or if the switch fails to close. Using the symbols in Figure 4.1, a fault tree for the example is shown in Figure 4.2. The single capital letters in the diagram of Figure 4.2 denote corresponding fault events (e.g., *T*: Dark room [top fault event]).

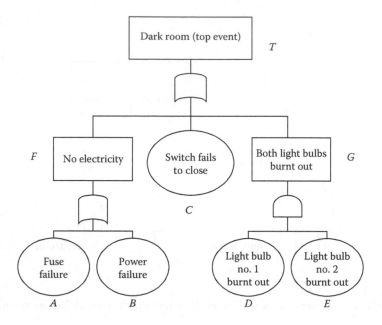

FIGURE 4.2
A fault tree for the top event: Dark room.

4.2.1 Probability Evaluation of Fault Trees

When the probabilities of occurrence of primary/basic fault events are known, then the probability of occurrence of the top fault event can be calculated. This can only be calculated by first calculating the probabilities of occurrence of the output fault events of all the lower and intermediate logic gates (e.g., AND and OR gates).

Thus, the occurrence probability of the AND gate output fault event, say A, is given by Dhillon [1] as

$$P(A) = \prod_{i=1}^{m} P(A_i), \tag{4.1}$$

where
   $P(A)$ is the probability of occurrence of the AND gate output fault event A.
   m is the number of AND gate input independent fault events.
   $P(A_i)$ is the occurrence probability of the AND gate input fault event A_i; for $i = 1, 2, 3, \ldots, m$.

Similarly, the occurrence probability of the OR gate output fault event, *B*, is given by Dhillon [1] as

$$P(B) = 1 - \prod_{i=1}^{n} \left[1 - P(B_i) \right],$$ (4.2)

where

P(B) is the probability of occurrence of the OR gate output fault event *B*.

n is the number of OR gate input independent fault events.

P(B_i) is the occurrence probability of the OR gate input fault event *B_i*; for *i* = 1, 2, 3, ..., *n*.

Example 4.2

Assume that the probabilities of occurrence of fault events *A, B, C, D,* and *E* in Figure 4.2 are 0.02, 0.01, 0.04, 0.08, and 0.09, respectively. Calculate the occurrence probability of the top fault event *T* (dark room) with the aid of Equations 4.1 and 4.2.

By inserting the specified occurrence probability values of fault events *A* and *B* into Equation 4.2, we obtain

$$P(F) = 1 - [(1 - 0.02)(1 - 0.01)] = 0.0298,$$

where

P(F) is the occurrence probability of fault event *F* (no electricity).

Similarly, by inserting the specified occurrence probability values of fault events *D* and *E* into Equation 4.1, we obtain

$$P(G) = (0.08)(0.09) = 0.0072,$$

where

P(G) is the occurrence probability of fault event *G* (both light bulbs burnt out).

By inserting the specified data value and the two calculated values earlier into Equation 4.2, we obtain

$$P(T) = 1 - [(1 - 0.04)(1 - 0.0298)(1 - 0.0072)] = 0.0753,$$

where

P(T) is the occurrence probability of fault event *T* (dark room).

Thus, the occurrence probability of the top fault event *T* (dark room) is 0.0753.

4.2.2 FTA Advantages and Disadvantages

There are many advantages and disadvantages of the FTA. Some of its advantages are as follows [1,8]:

- Useful to identify failures deductively
- Provides insight into the system behavior
- Useful to handle complex systems more easily
- Serves as a graphic aid for system management
- Useful in providing options for management and others for performing either quantitative or qualitative reliability analysis
- Allows concentration on one specific failure at a time
- Requires the analyst to thoroughly comprehend the system under consideration before starting the analysis

In contrast, some of the disadvantages of the FTA are as follows [1,8]:

- It is a time-consuming and costly method.
- Results are quite difficult to check.
- It considers components/parts in either an operational state or a failed state (i.e., partial failure states of the components/parts are difficult to handle).

4.3 Markov Method

The Markov method is widely used for performing reliability-related analysis of engineering systems and is named after a Russian mathematician, Andrei A. Markov (1856–1922). The method is commonly used for modeling repairable systems with constant failure and repair rates. The following three assumptions are associated with this method [9]:

- The transitional probability from one system state to another in the finite time interval Δt is given by $\alpha \Delta t$, where α is the transition rate (e.g., system failure or repair rate) from one system state to another.
- The probability of more than one transition occurrence in the finite time interval Δt from one system state to another is negligible (e.g., $(\alpha \Delta t)(\alpha \Delta t) \to 0$).
- All occurrences are independent of each other.

The application of this method is demonstrated by solving the following example:

Example 4.3

Assume that an engineering system can be in either a working or a failed state. The system constant failure and repair rates are λ_{es} and μ_{es}, respectively. The engineering system state space diagram is shown in Figure 4.3. The numerals in circles denote the engineering system states. Develop equations for the engineering system time-dependent and steady-state availabilities and unavailabilities, reliability, and mean time to failure with the aid of the Markov method.

With the aid of the Markov method, we write down the following equations for the engineering system states 0 and 1 shown in Figure 4.3, respectively:

$$P_0(t + \Delta t) = P_0(t)(1 - \lambda_{es}\Delta t) + P_1(t)\mu_{es}\Delta t, \tag{4.3}$$

$$P_1(t + \Delta t) = P_1(t)(1 - \mu_{es}\Delta t) + P_0(t)\lambda_{es}\Delta t, \tag{4.4}$$

where
t is time.
$P_0(t + \Delta t)$ is the probability of the engineering system being in working state 0 at time $(t + \Delta t)$.
$P_1(t + \Delta t)$ is the probability of the engineering system being in failed state 1 at time $(t + \Delta t)$.
$\lambda_{es}\Delta t$ is the probability of engineering system failure in finite time interval Δt.
$\mu_{es}\Delta t$ is the probability of engineering system repair in finite time interval Δt.
$(1 - \lambda_{es}\Delta t)$ is the probability of no failure in finite time interval Δt.
$(1 - \mu_{es}\Delta t)$ is the probability of no repair in finite time interval Δt.
$P_j(t)$ is the probability that the engineering system is in state j at time t, for $j = 0, 1$.

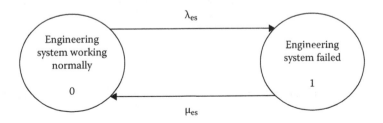

FIGURE 4.3
Engineering system state space diagram.

From Equation 4.3, we obtain

$$P_0(t + \Delta t) = P_0(t) - P_0(t)\lambda_{es}\Delta t + P_1(t)\mu_{es}\Delta t. \tag{4.5}$$

From Equation 4.5, we write

$$\lim_{\Delta t \to 0} \frac{P_0(t + \Delta t) - P_0(t)}{\Delta t} = -P_0(t)\lambda_{es} + P_1(t)\mu_{es}. \tag{4.6}$$

Thus, from Equation 4.6, we get

$$\frac{dP_0(t)}{dt} + P_0(t)\lambda_{es} = P_1(t)\mu_{es}. \tag{4.7}$$

Similarly, using Equation 4.4, we obtain

$$\frac{dP_1(t)}{dt} + P_1(t)\mu_{es} = P_0(t)\lambda_{es}. \tag{4.8}$$

At time $t = 0$, $P_0(0) = 1$ and $P_1(0) = 0$.

By solving Equations 4.7 and 4.8, we get [1] the following:

$$P_0(t) = \frac{\mu_{es}}{\left(\lambda_{es} + \mu_{es}\right)} + \frac{\lambda_{es}}{\left(\lambda_{es} + \mu_{es}\right)} e^{-(\lambda_{es} + \mu_{es})t}, \tag{4.9}$$

$$P_1(t) = \frac{\lambda_{es}}{\left(\lambda_{es} + \mu_{es}\right)} - \frac{\lambda_{es}}{\left(\lambda_{es} + \mu_{es}\right)} e^{-(\lambda_{es} + \mu_{es})t}. \tag{4.10}$$

Thus, the engineering system time-dependent availability and unavailability, respectively, are

$$AV_{es}(t) = P_0(t) = \frac{\mu_{es}}{\lambda_{es} + \mu_{es}} + \frac{\lambda_{es}}{\lambda_{es} + \mu_{es}} e^{-(\lambda_{es} + \mu_{es})t} \tag{4.11}$$

and

$$UA_{es}(t) = P_1(t) = \frac{\lambda_{es}}{\left(\lambda_{es} + \mu_{es}\right)} - \frac{\lambda_{es}}{\left(\lambda_{es} + \mu_{es}\right)} e^{-(\lambda_{es} + \mu_{es})t}, \tag{4.12}$$

where

$AV_{es}(t)$ is the engineering system time-dependent availability.

$UA_{es}(t)$ is the engineering system time-dependent unavailability.

By letting time t go to infinity in Equations 4.11 and 4.12, we obtain [1]

$$AV_{es} = \lim_{t \to \infty} AV_{es}(t) = \frac{\mu_{es}}{\lambda_{es} + \mu_{es}} \tag{4.13}$$

and

$$UA_{es} = \lim_{t \to \infty} UA_{es}(t) = \frac{\lambda_{es}}{\lambda_{es} + \mu_{es}}, \tag{4.14}$$

where
AV_{es} is the engineering system steady-state availability.
UA_{es} is the engineering system steady-state unavailability.

For $\mu_{es} = 0$, from Equation 4.9, we obtain

$$R_{es}(t) = P_0(t) = e^{-\lambda_{es}t}, \tag{4.15}$$

where
$R_{es}(t)$ is the engineering system reliability at time t.

By integrating Equation 4.15 over the time interval [0, ∞], we obtain the following equation for the engineering system mean time to failure [1]:

$$MTTF_{es} = \int_0^\infty e^{-\lambda_{es}t}\, dt$$
$$= \frac{1}{\lambda_{es}}, \tag{4.16}$$

where
$MTTF_{es}$ is the engineering system mean time to failure.

Thus, the engineering system time-dependent and steady-state availabilities and unavailabilities, reliability, and mean time to failure are given by Equations 4.11, 4.13, 4.12, 4.14, 4.15, and 4.16, respectively.

Example 4.4

Assume that the constant failure and repair rates of an engineering system are 0.006 failures/hour and 0.008 repairs/hour, respectively. Calculate the engineering system steady-state unavailability and unavailability during a 50-hour mission.

By inserting the specified data values into Equations 4.14 and 4.12, we obtain

$$UA_{es} = \frac{0.006}{0.006 + 0.008} = 0.4285$$

and

$$UA_{es}(50) = \frac{0.006}{(0.006+0.008)} - \frac{0.006}{(0.006+0.008)} e^{-(0.006+0.008)(50)}$$

$$= 0.2157.$$

Thus, the engineering system steady-state unavailability and unavailability during a 50-hour mission are 0.4285 and 0.2157, respectively.

4.4 Failure Modes and Effect Analysis

Failure modes and effect analysis (FMEA) is a widely used method for analyzing the reliability of engineering systems. It can simply be described as an approach to analyze the effects of potential failure modes in the system [1,10]. The history of this method goes back to the early 1950s with the development of flight control systems, when the US Navy's Bureau of Aeronautics developed a requirement called *failure analysis* for establishing a mechanism for reliability control over the detail design effort [11].

Subsequently, the term *failure analysis* was changed over to *failure modes and effect analysis*. Generally, the following seven steps are followed to perform FMEA [3,8]:

- *Step 1:* Define system boundaries and its associated requirements.
- *Step 2:* List system subsystems and components.
- *Step 3:* List each component's failure modes, the description, and the identification.
- *Step 4:* Assign failure occurrence probability/rate to each component failure mode.
- *Step 5:* List each failure mode effect/effects on subsystem(s), system, and plant.
- *Step 6:* Enter necessary remarks for each failure mode.
- *Step 7:* Review each critical failure mode and take necessary actions.

Prior to the implementation of FMEA, there are a number of factors that must be explored. Four of these factors are presented in the following [8,12,13]:

- Making appropriate decisions based on the risk priority number
- Examination of each conceivable failure mode by all the involved professionals

- Obtaining involved engineer's approval and support
- Measuring benefits/costs

Over the years, professionals involved in reliability analysis have established a number of guidelines/facts concerning FMEA. Four of these guidelines/facts are as follows [13]:

- FMEA is not a method for selecting the optimum design concept.
- Developing the most of FMEA in a meeting should be avoided.
- FMEA is not designed for superseding the engineer's work.
- FMEA has certain limitations.

Some of the main benefits of performing FMEA are presented in the following [1,8,12,13]:

- A useful method for comparing designs and highlighting safety concerns
- A systematic approach for classifying/categorizing hardware failures
- A useful method for improving communication between design interface personnel
- A useful method that starts from the detailed level and works upward
- A useful approach for safeguarding against repeating the same mistakes in the future
- A useful visibility tool for management that reduces product development cost and time
- A useful method for reducing engineering changes and for improving the efficiency of test planning
- A helpful approach for understanding and improving customer satisfaction

4.5 Probability Tree Analysis

Probability tree analysis is a method used to perform task analysis by diagrammatically representing critical human-related actions and other events concerning the system under consideration. Frequently, the probability tree method is used for performing task analysis in the technique for human error rate prediction [2,3,14]. In this method, the diagrammatic task analysis is represented by the probability tree branches. More specifically, the branching limbs of the tree denote the outcomes (i.e., success or failure) of each event, and each branch is assigned a value for the occurrence probability.

Some of the advantages of the method are as follows [2,14,15]:

- A quite useful visibility tool
- Decreases the error occurrence probabilities in computation because of simplification in the computational process
- Incorporates, with some modifications, factors such as emotional stress, interaction effects, and interaction stress

The application of the method is demonstrated through the two examples presented in the following.

Example 4.5

Assume that an engineering system maintenance person performs two independent maintenance-related tasks: a and b. Each of these two tasks can be carried out either correctly or incorrectly, and task a is performed before task b.

Develop a probability tree and obtain expressions for the probability of successfully and the probability of not successfully accomplishing the overall mission by the engineering system maintenance person.

In this case, the engineering system maintenance person performs task a correctly or incorrectly and then proceeds to carrying out task b. Task b can also be carried out either correctly or incorrectly by the engineering system maintenance person. This entire scenario is shown in Figure 4.4.

The symbols used in the Figure 4.4 diagram are defined in the following:

- a is task a performed correctly.
- b is task b performed correctly.
- $\bar{a}$ is task a performed incorrectly.
- $\bar{b}$ is task b performed incorrectly.

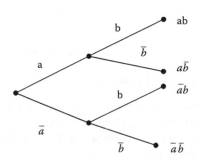

FIGURE 4.4
Probability tree diagram for the engineering system maintenance person carrying out tasks a and b.

By examining the Figure 4.4 diagram, it is concluded that there is only one possibility (i.e., ab) for successfully accomplishing the overall mission by the engineering system maintenance person.

Thus, the probability of successfully accomplishing the overall mission by the engineering system maintenance person is given by

$$P_s = P(ab) = P_a P_b, \tag{4.17}$$

where

P_s is the probability of successfully accomplishing the overall mission by the engineering system maintenance person.

P_a is the probability of performing independent task a correctly by the engineering system maintenance person.

P_b is the probability of performing independent task b correctly by the engineering system maintenance person.

Similarly, by examining the Figure 4.4 diagram, it is concluded that there are three possibilities (i.e., $a\bar{b}$, $\bar{a}b$, and $\bar{a}\bar{b}$) for not successfully accomplishing the overall mission by the engineering system maintenance person. Thus, the probability of not successfully accomplishing the overall mission by the engineering system maintenance person is given by

$$P_f = P(a\bar{b}) + P(\bar{a}b) + P(\bar{a}\bar{b})$$
$$= P_a P_{\bar{b}} + P_{\bar{a}} P_b + P_{\bar{a}} P_{\bar{b}}, \tag{4.18}$$

where

P_f is the probability of not successfully accomplishing the overall mission by the engineering system maintenance person.

$P_{\bar{a}}$ is the probability of performing independent task a incorrectly by the engineering system maintenance person.

$P_{\bar{b}}$ is the probability of performing independent task b incorrectly by the engineering system maintenance person.

Thus, Equations 4.17 and 4.18 are the expressions for the probability of successfully and the probability of not successfully accomplishing the overall mission, respectively, by the engineering system maintenance person.

Example 4.6

Calculate the probability of successfully and the probability of not successfully accomplishing the overall mission by the engineering system maintenance person, if the probabilities of correctly performing tasks a and b are 0.90 and 0.98, respectively.

By substituting the given data values into Equation 4.17, we obtain

$$P_s = (0.90)(0.98) = 0.882.$$

Similarly, by substituting the given data values into Equation 4.18, we obtain

$$P_f = P_a P_{\bar{b}} + P_{\bar{a}} P_b + P_{\bar{a}\bar{b}}$$

$$= P_a(1 - P_b) + (1 - P_a)P_b + (1 - P_a)(1 - P_b)$$

$$= (0.90)(1 - 0.98) + (1 - 0.90)(0.98) + (1 - 0.90)(1 - 0.98)$$

$$= 0.118.$$

Thus, the probability of successfully and the probability of not successfully accomplishing the overall mission by the engineering system maintenance person are 0.882 and 0.118, respectively.

4.6 Technique of Operation Review

Technique of operation review (TOR) is a method developed in the early 1970s by D.A. Weaver of the American Society of Safety Engineers, and it seeks to highlight systemic causes for an adverse event rather than assigning blame with respect to safety [5,8,16]. The method permits management personnel and workers to work jointly to analyze workplace-related failures, accidents, and incidents. Thus, TOR may simply be described as a hands-on analytical method for highlighting the root system causes of an operation malfunction/failure [16].

TOR uses a worksheet containing simple terms that require yes/no decisions and is activated by an adverse incident occurring at a certain point in time and location that involves certain persons. It is to be noted that this method is not a hypothetical process and demands a clear systematic evaluation of the circumstances surrounding the incident/accident under consideration. Ultimately, TOR highlights how the organization/company could have prevented the occurrence of accident/incident.

The method is composed of the eight steps presented in the following [5,8,17]:

- *Step 1:* Form the TOR team with members belonging to all concerned areas.
- *Step 2:* Hold a roundtable session to impart common knowledge to all members of the TOR team.
- *Step 3:* Highlight one key systemic factor that played a pivotal role in the occurrence of the incident/accident. This factor must be based on team members' consensus and serves as an important starting point for further investigation.

- *Step 4:* Use the team consensus when responding to a sequence of yes/no options.
- *Step 5:* Evaluate the highlighted factors, ensuring that there is a clear-cut consensus among the members of the team with respect to the assessment of each and every factor.
- *Step 6:* Prioritize the contributory factors by starting with the most serious one.
- *Step 7:* Develop necessary preventive/corrective strategies with respect to each and every contributory factor.
- *Step 8:* Conduct the implementation of the strategies.

Finally, it is added that the main strength of this method is the involvement of line personnel in the analysis process. In contrast, the method's main weakness is that it is an after-the-fact process.

4.7 Hazard and Operability Analysis

Hazard and operability analysis (HAZOP) is a systematic approach for highlighting hazards and operating-related problems in a facility. It has proven to be a very useful tool for highlighting unforeseen hazards designed into facilities due to various reasons or introduced into already existing facilities due to factors such as changes carried out to process-related conditions or operating procedures.

The basic objectives of HAZOP are as follows [5,8,18]:

- To decide whether deviations from design intentions can result in operating-related problems/hazards
- To review each process/facility part to discover how deviations from the design intentions can take place
- To develop a complete facility/process description

A HAZOP study can be conducted in five steps as shown in Figure 4.5 [8,18].

Step 1 (i.e., establish study scope and objectives) is concerned with developing the study scope and the objectives by taking into consideration all relevant factors. Step 2 (i.e., form HAZOP team) is concerned with forming a HAZOP team by ensuring that the team is composed of persons from the area of design and operation with the appropriate experience for determining the effects of deviations from the intended application.

Step 3 (i.e., collect relevant information) is concerned with obtaining the necessary documentation; process description; and drawings, including items

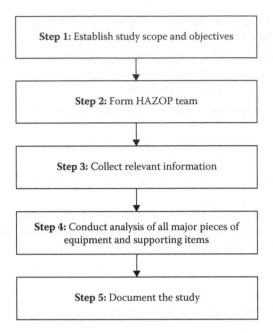

FIGURE 4.5
Steps for conducting HAZOP study.

such as operating and maintenance procedures, layout drawings, process control logic diagrams, equipment specifications, emergency response procedures, and process flow sheets. Step 4 (i.e., conduct analysis of all major pieces of equipment and supporting items) is concerned with analyzing all the major items of equipment and all the supporting equipment, piping, and instrumentation with the aid of step 3 documents.

Finally, step 5 (i.e., document the study) is concerned with documenting the consequences of any deviation from the norm as well as a deviations' summary from the norm and summary of those deviations considered hazardous and credible.

4.8 Interface Safety Analysis

Interface safety analysis (ISA) is a method concerned with determining the incompatibilities between subsystems and assemblies of equipment/product that could lead to accidents. The method establishes that distinct units/parts can be integrated into a viable system and that an individual unit's or part's normal operation will not impair the performance of or damage another unit/part or the whole system/equipment. Although ISA considers various relationships, they can be grouped under three classifications: flow relationships, physical relationships, and functional relationships. Each of these classifications is discussed in the following, separately [5,8,19].

4.8.1 Classification I: Flow Relationships

Flow relationships are concerned with two or more units/items. For example, the flow between two units/items may involve electrical energy, fuel, air, steam, water, or lubricating oil. Furthermore, the flow could be unconfined, such as heat radiation from one item to another item. The problems frequently experienced with many products include the proper flow of fluids and energy from one unit to another unit through confined passages, consequently leading to safety-related problems.

The causes of flow-associated problems include faulty connections between items/units and partial or complete interconnection failure. In the case of fluids, from the safety perspective, the factors that must be considered with utmost care include contamination, flammability, toxicity, odor, loss of pressure, and lubricity.

4.8.2 Classification II: Physical Relationships

Physical relationships are concerned with the physical aspects of units/products. For example, two units/products might be well designed and manufactured and individually operate quite effectively, but they may experience difficulties in fitting together properly because of dimension-related differences, or there may be other incompatibilities that may result in safety problems. Some examples of the other problems are as follows:

- Restricted/impossible access to or egress from equipment
- A very small clearance between the units; thus, the units may be damaged during the removal activities
- Impossible to join, mate, or tighten parts properly

4.8.3 Classification III: Functional Relationships

Functional relationships are concerned with multiple units/items. For example, in a situation where a unit's outputs constitute the inputs to the downstream unit(s), any error in outputs and inputs may cause damage to the downstream unit(s), thereby creating a safety-related problem. Such outputs could be in conditions such as excessive outputs, unprogrammed outputs, degraded outputs, erratic outputs, and zero outputs.

4.9 Maintenance Program Effectiveness Evaluation Approach for Managers

In the 1970s, the US Energy Research and Development Administration conducted a study on engineering maintenance management-related matters [20]. As the result of this study, an approach for evaluating the effectiveness of an ongoing maintenance program was developed. The approach is composed of the following 10 questions for maintenance managers to self-evaluate their ongoing maintenance effort [20,21]:

- Are you aware of whether safety practices are being followed properly?
- Are you aware of how your craft persons spend their time, i.e., delays, travel, etc.?
- In regard to job costs, are you in a good position for comparing the *should* with the *what*?
- Are you fully aware of how much time your supervisor spends at the desk and at the job site?
- Are you fully aware of what equipment/facility and activity consume most of the maintenance money?
- Have you balanced your spare parts inventory in regard to carrying cost against anticipated downtime losses?
- Are you providing the craft persons with the right quantity and the quality of material when and where they need it?
- Are you fully aware if the craft persons use correct tools and methods for performing their tasks?
- Do you have an effective base for performing productivity-related measurements? Is productivity improving?
- Do you ensure that all maintainability-related factors are considered correctly during the design of new or modified facility/equipment?

If an unqualified *yes* is the answer to each of the above 10 questions, then your ongoing maintenance program is on a sound footing to satisfy organization-related objectives. Otherwise, appropriate corrective measures are needed.

4.10 Indices for Maintenance Management Analysis

In the engineering industrial sector, the management uses various approaches for measuring effectiveness of the maintenance activity concerning engineering systems. Often, it uses various types of indices to manage and control such maintenance activity. The basic objective of these indices is to encourage maintenance management personnel for improving on past performance.

This section presents a number of such indices divided into two categories: broad indices and specific indices [7,21–24]. The broad indices category indicates the overall performance of the organization/facility in regard to the maintenance activity, and the specific indices category indicates the performance in specific areas of the maintenance activity. The values of all these indices are plotted periodically to show the trends. Both these categories of indices are presented in the following, separately.

4.10.1 Category I: Broad Indices

Category I: Broad indices contains the following three indices:

- *Index I*

    Index I is defined by

$$\theta_1 = \frac{TC_m}{TI_{pe}}, \tag{4.19}$$

    where
    θ_1 is the index parameter.
    TC_m is the total maintenance cost.
    TI_{pe} is the total investment in plant and equipment.

    This index relates the total maintenance-related cost to the total investment in equipment and plant. In the chemical and steel industrial sectors, the approximate average figures for θ_1 are 3.8% and 8.6%, respectively.

- *Index II*

    Index II is defined by

$$\theta_2 = \frac{TC_m}{O_t}, \tag{4.20}$$

    where
    θ_2 is the index parameter.
    O_t is the total output expressed in tons, megawatts, gallons, etc.

    This index relates the total maintenance-related cost to the total output by the facility/organization.

- *Index III*

  Index III is expressed by

$$\theta_3 = \frac{TC_m}{S_t} \tag{4.21}$$

where
  θ_3 is the index parameter.
  S_t is the total sales.

Experiences over the years indicate that the average expenditure for maintenance activity for all industrial sectors was about 5% of sales. However, there was a wide variation among industrial sectors. For example, the average values of θ_3 for chemical and steel industrial sectors were 6.8% and 12.8%, respectively.

4.10.2 Category II: Specific Indices

Category II: Specific indices contains the following 12 indices:

- *Index I*

  Index I can be used to measure maintenance effectiveness and is expressed by

$$\alpha_1 = \frac{MH_{ue}}{MH_{tm}}, \tag{4.22}$$

where
  α_1 is the index parameter.
  MH_{tm} is the total maintenance human-hours worked.
  MH_{ue} is the human-hours of unscheduled and emergency jobs.

- *Index II*

  Index II is quite useful for controlling the preventive maintenance activity within a maintenance organization/facility and is expressed by

$$\alpha_2 = \frac{TT_{pm}}{TT_{em}}, \tag{4.23}$$

where
  α_2 is the index parameter.
  TT_{em} is the total time spent for the entire maintenance activity.
  TT_{pm} is the total time spent in performing preventive maintenance.

- *Index III*

  Index III is concerned with measuring inspection effectiveness and is expressed by

  $$\alpha_3 = \frac{J_{ri}}{I_{tc}}, \tag{4.24}$$

  where
  α_3 is the index parameter.
  I_{tc} is the total number of inspections completed.
  J_{ri} is the number of jobs resulting from inspections.

- *Index IV*

  Index IV can be used to measure maintenance and is defined by

  $$\alpha_4 = \frac{DT_{cb}}{DT_t}, \tag{4.25}$$

  where
  α_4 is the index parameter.
  DT_t is the total downtime.
  DT_{cb} is the downtime caused by breakdowns.

- *Index V*

  Index V is concerned with maintenance overhead control and is defined by

  $$\alpha_5 = \frac{AC_m}{TC_m}, \tag{4.26}$$

  where
  α_5 is the index parameter.
  TC_m is the total maintenance cost.
  AC_m is the total maintenance administration cost.

- *Index VI*

  Index VI can be used to measure the accuracy of the maintenance budget plan and is defined by

  $$\alpha_6 = \frac{AC_{tm}}{TBC_m}, \tag{4.27}$$

where
  α_6 is the index parameter.
  TBC_m is the total budgeted maintenance cost.
  AC_{tm} is the total actual maintenance cost.

- *Index VII*

  Index VII is quite useful in scheduling work and is defined by

  $$\alpha_7 = \frac{PJ_c}{PJ_{tn}},\qquad(4.28)$$

where
  α_7 is the index parameter.
  PJ_{tn} is the total number of planned jobs.
  PJ_c is the total number of planned jobs completed by established due dates.

- *Index VIII*

  Index VIII relates maintenance-related cost to manufacturing cost and is expressed by

  $$\alpha_8 = \frac{TC_m}{TMC},\qquad(4.29)$$

where
  α_8 is the index parameter.
  TMC is the total manufacturing cost.
  TC_m is the total maintenance cost.

- *Index IX*

  Index IX is quite useful for monitoring the progress in cost reduction-related efforts and is expressed by

  $$\alpha_9 = \frac{PMH_{sj}}{MC_{up}},\qquad(4.30)$$

where
  α_9 is the index parameter.
  MC_{up} is the maintenance cost per unit of production.
  PMH_{sj} is the percentage of maintenance human-hours spent on scheduled jobs.

- *Index X*

  Index X is quite useful in the area of material control and is expressed by

$$\alpha_{10} = \frac{PJAM_{tn}}{PJ_{tn}},$$

(4.31)

where

α_{10} is the index parameter.
PJ_{tn} is the total number of planned jobs.
$PJAM_{tn}$ is the total number of planned jobs awaiting material.

- *Index XI*

  Index XI relates maintenance-related cost to human-hours worked and is defined by

$$\alpha_{11} = \frac{TC_m}{MH_{tw}},$$

(4.32)

where

α_{11} is the index parameter.
MH_{tw} is the total number of human-hours worked.
TC_m is the total maintenance cost.

- *Index XII*

  Index XII relates to maintenance materials and labor-related costs and is defined by

$$\alpha_{12} = \frac{LC_{tm}}{MC_{tm}},$$

(4.33)

where

α_{12} is the index parameter.
MC_{tm} is the total maintenance material cost.
LC_{tm} is the total maintenance labor cost.

PROBLEMS

1. What are the main prerequisites associated with FTA?
2. What are the advantages and the disadvantages of FTA?

3. Assume that a windowless room contains one switch and four light bulbs. Develop a fault tree for the top (undesired) fault event *dark room* if the switch can only fail to close.

4. What are the assumptions associated with the Markov method?

5. Prove Equations 4.11 and 4.12 by using Equations 4.7 and 4.8.

6. Compare fault FTA with probability tree analysis.

7. Describe the TOR.

8. Compare HAZOP with ISA.

9. Describe the maintenance program effectiveness evaluation approach for managers.

10. Describe the indices for maintenance management analysis and define at least four of such indices.

References

1. Dhillon, B.S., *Design Reliability: Fundamentals, and Applications*, CRC Press, Boca Raton, FL, 1999.
2. Dhillon, B.S., Singh, C., *Engineering Reliability: New Techniques and Applications*, Wiley, New York, 1981.
3. Dhillon, B.S., *Human Reliability: With Human Factors*, Pergamon Press, New York, 1986.
4. Hammer, W., Price, D., *Occupational Safety Management and Engineering*, Prentice Hall, Upper Saddle River, NJ, 2001.
5. Dhillon, B.S., *Engineering Safety: Fundamentals, Techniques, and Applications*, World Scientific Publishing, River Edge, NJ, 2003.
6. Higgins, L.R., Editor, *Maintenance Engineering Handbook*, McGraw-Hill Book Company, New York, 1988.
7. Stoneham, D., *The Maintenance Management and Technology Handbook*, Elsevier Science, Oxford, 1998.
8. Dhillon, B.S., *Transportation Systems Reliability and Safety*, CRC Press, Boca Raton, FL, 2011.
9. Shooman, M.L., *Probabilistic Reliability: An Engineering Approach*, McGraw-Hill Book Company, New York, 1968.
10. Omdahl, T.P., Editor, *Reliability, Availability, and Maintainability (RAM) Dictionary*, American Society for Quality Control (ASQC) Press, Milwaukee, WI, 1988.
11. MIL-F-18372 (Aer), *General Specification for Design, Installation, and Test of Aircraft Flight Control Systems*, Bureau of Naval Weapons, Department of the Navy, Washington, DC.
12. McDermott, R.E., Mikulak, K.J., Beauregard, M.R., *The Basics of FMEA*, Quality Resources, New York, 1996.

13. Palady, P., *Failure Modes and Effects Analysis*, PT Publications, West Palm Beach, FL, 1995.
14. Dhillon, B.S., *Robot System Reliability and Safety: A Modern Approach*, CRC Press, Boca Raton, FL, 2015.
15. Swain, A.D., An error-cause removal program for industry, *Human Factors*, Vol. 12, 1973, pp. 207–221.
16. Hallock, R.G., Technique of operations review analysis: Determines cause of accident/incident, *Safety and Health*, Vol. 60, No. 8, 1991, pp. 38–39.
17. Goetsch, D.L., *Occupational Safety and Health*, Prentice Hall, Englewood Cliffs, NJ, 1996.
18. CAN/CSA-Q6340-91, *Risk Analysis Requirements and Guidelines*, Canadian Standards Association (CSA), Rexdale, ON, 1991.
19. Hammer, W., *Product Safety Management and Engineering*, Prentice Hall, Englewood Cliffs, NJ, 1980.
20. ERHQ-0004, *Maintenance Manager's Guide*, Energy Research and Development Administration, Washington, DC, 1976.
21. Dhillon, B.S., *Engineering Maintenance: A Modern Approach*, CRC Press, Boca Raton, FL, 2002.
22. Niebel, B.W., *Engineering Maintenance Management*, Marcel Dekker, New York, 1994.
23. Hartmann, E., Knapp, D.J., Johnstone, J.J., Ward, K.G., *How to Manage Maintenance*, American Management Association, New York, 1994.
24. Westerkamp, T.A., *Maintenance Manager's Standard Manual*, Prentice Hall, Paramus, NJ, 1997.

5

Computer, Internet, and Robot System Reliability

5.1 Introduction

Over the years, computer applications have increased at an alarming rate, ranging from those for personal use to those controlling various types of sophisticated systems. As computer failures can, directly or indirectly, affect our day-to-day life, their reliability has become an important issue to the population at large. Furthermore, the reliability of computer systems used in areas such as nuclear power generation, aerospace, and defense is of utmost importance because their failures could be very costly and catastrophic.

The history of the Internet may be traced back to the late 1960s with the development of the Advanced Research Projects Agency Network [1]. The Internet has grown from only 4 hosts in 1969 to about 38 million sites and 147 million hosts in 2002, and nowadays billions of people around the globe use the services of the Internet [1,2]. Today, Internet reliability has become very important to the global economy and other areas, because Internet failures can result in millions of dollars in losses and interrupt the day-to-day routines of its vast number of end users around the globe [3].

Nowadays, robots are widely used to perform various types of tasks in the industrial sector. As many different types of parts (e.g., electronic, hydraulic, pneumatic, and mechanical) are used in robots, this makes the task of producing highly reliable robots very challenging and time consuming. Needless to say, robot reliability has become an important issue.

This chapter presents various important aspects of computer, Internet, and robot system reliability.

5.2 Computer System Reliability Issue-Related Factors and Computer Failure Sources

As there are many issues concerned with computer system reliability, some of the important factors to consider are as follows [4–6]:

- The logic elements are the main components/parts of computers that have troublesome reliability-related features. The proper determination of the reliability of such elements is impossible in many situations, and their defects cannot be effectively healed.

- Computer-related failures are highly varied in character. For example, a computer system component or part may fail permanently or it may experience a transient fault due to its environment.

- For fault tolerance, modern computers are composed of redundancy schemes; and although advances made over the years have brought various types of improvements, there are still many practical and theoretical difficulties that remain to be effectively overcome.

- It could be very difficult to detect hardware design errors at the lowest system levels prior to the production and installation phases. Therefore, it is possible that hardware design errors may lead to situations where it is impossible to distinguish operation errors due to such oversights from the ones due to transient physical faults.

- Usually, dynamic fault tolerance is the most powerful type of self-repair in computers but is quite difficult to analyze. Nevertheless, for certain applications, it is very important and it cannot simply be overlooked.

There are many sources that lead to computer failures. The eight major such sources are shown in Figure 5.1 [6–9].

Six of the sources shown in Figure 5.1 are described in the following.

Communication network failures are concerned with intermodule communication, and most of these failures are usually of a transient nature. The application of *vertical parity* logic can help to detect around two-thirds of errors in communication lines. Peripheral device failures are important, but they rarely cause a system shutdown. The frequently occurring errors in peripheral devices are transient or intermittent, and the usual reason for their occurrence is the peripheral devices' electromechanical nature. Human errors usually occur due to operator oversights and mistakes. Frequently, operator errors take place during starting up, running, and shutting down the computer system.

Environmental failures occur due to causes such as air conditioning equipment failure, fires, electromagnetic interference, and earthquakes. In the case of power failures, factors such as total power loss from the local

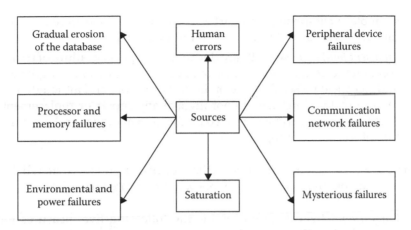

FIGURE 5.1
Major sources of computer failures.

utility company and transient fluctuations in frequency or voltage are the causes of their occurrence. Mysterious failures occur unexpectedly; thus, in real-life systems, such failures are never classified or categorized properly. For example, when a normally operating system stops operating at once suddenly without indicating any problem (i.e., software, hardware, etc.), the failure is known as a *mysterious failure*. Processor failures/errors are catastrophic, but their occurrence is rare, as there are times when the central processor malfunctions to execute instructions correctly due to a dropped bit. Nowadays, the occurrence of memory purity errors is rare because of improvements in hardware reliability, and they are not necessarily fatal.

5.3 Computer-Related Fault Classifications and Reliability Measures

Generally, for computer system reliability modeling and evaluation, an effective method to classify computer-related faults is on the basis of their duration. Thus, the computer-related faults may be classified under two categories [6,10]:

- *Permanent faults:* These faults are often due to catastrophic failures of parts/components. In this situation, the components' or the parts' failures are irreversible and permanent and require repair or replacement. These faults have a failure rate that depends on the surrounding environment and are characterized by long duration. For example, a component or a part will usually have a different failure rate in power-on and power-off conditions [6,11].

- *Transient faults:* These faults are due to the temporary failure of parts/ components or the external interference such as power dips, glitches, and electrical noise. They are of limited duration, and although they require restoration, they do not involve any repair or replacement. This type of fault is characterized by the arrival modes and the duration of transients [6,10].

There are various measures used in the area of computer system reliability, and they may be divided under the following two categories [5,6]:

- *Category I:* This category contains the following five measures for gracefully handling degrading systems:
  - *Mean computation before failure:* This is the expected amount of computation available on the system before failure.
  - *Capacity threshold:* This is the time at which a specific value of computation availability is reached.
  - *Computation reliability:* This is the probability that the system will, without an error, execute a task of length, say, x that began at time t.
  - *Computation availability:* This is the system expected computation capacity at a specified time t.
  - *Computation threshold:* This is the time at which a specific value of computation reliability is reached for task whose length is, say, x.
- *Category II:* This category contains those measures that are considered suitable for configurations such as hybrid, standby, and massively redundant systems. The measures are system reliability, mission time, system availability, and MTTF. It is to be noted that these measures may not be sufficient for gracefully evaluating degrading systems.

5.4 Fault Masking

The term *fault masking*, in the area of fault-tolerant computing, is used in the sense that a system with redundancy can tolerate a number of failures prior to its own failure. Thus, the implication of the term *masking* is that some kind of problem has occurred somewhere within a digital system, but because of design, the problem does not affect the system's overall operation. The best-known fault masking method is probably modular redundancy.

5.4.1 Triple Modular Redundancy

In the case of triple modular redundancy (TMR), three identical modules/units perform the same task simultaneously and a voter compares the modules' or the units' outputs and sides with the majority. The TMR system fails when at least two modules/units fail or the voter fails. It means that the TMR system can tolerate a single module/unit failure. Figure 5.2 shows the block diagram of the TMR system with voter [5,6,9].

For independent module and voter units, the reliability of the TMR system with voter, shown in Figure 5.2, is expressed by the following equation [6,9]:

$$R_{sv} = (3R^2 - 2R^3)R_v, \tag{5.1}$$

where
R_v is the voter unit reliability.
R is the module/unit reliability.
R_{sv} is the TMR system with voter reliability.

For constant failure rates of the TMR system modules/units and the voter unit and with the aid of the material presented in Chapter 3 and Equation 5.1, we get

$$R_{sv}(t) = (3e^{-2\lambda_m t} - e^{-3\lambda_m t})e^{-\lambda_m t}$$
$$= 3e^{-(2\lambda_m + \lambda_v)t} - 2e^{(3\lambda_m + \lambda_v)t}, \tag{5.2}$$

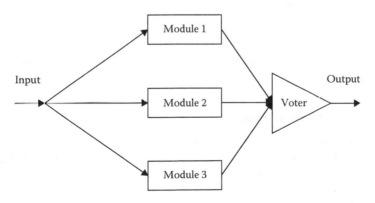

FIGURE 5.2
TMR system block diagram with voter.

where

$R_{sv}(t)$ is the TMR system with voter reliability at time t.
λ_m is the module/unit constant failure rate.
λ_v is the voter unit constant failure rate.

By integrating Equation 5.2 over the time interval from 0 to ∞, we obtain the following equation for the TMR system with voter MTTF [6,9]:

$$MTTF_{sv} = \int_0^\infty [3e^{-(2\lambda_m+\lambda_v)t} - 2e^{-(3\lambda_m+\lambda_v)t}]dt$$

$$= \frac{3}{(2\lambda_m+\lambda_v)} - \frac{2}{(3\lambda_m+\lambda_v)}. \tag{5.3}$$

Example 5.1

Assume that the constant failure rate of a module/unit of a TMR system with voter is $\lambda_m = 0.0005$ failures/hour and the constant failure rate of the voter unit is $\lambda_v = 0.0002$ failures/hour. Calculate the TMR system MTTF and reliability for a 400-hour mission.

By inserting the specified data values into Equation 5.3, we get

$$MTTF_{sv} = \frac{3}{2(0.0005)+0.0002} - \frac{2}{3(0.0005)+0.0002}$$

$$= 1323.53 \text{ hours.}$$

Similarly, by inserting the specified data values into Equation 5.2, we get

$$R_{sv}(400) = 3e^{-[2(0.0005)+0.0002](400)} - 2e^{-[3(0.0005)+0.0002](400)}$$

$$= 0.8431.$$

Thus, the TMR system MTTF and reliability are 1323.53 hours and 0.8431, respectively.

5.4.2 N-Modular Redundancy

N-modular redundancy (NMR) is the general form of the TMR which contains N identical modules/units instead of only three modules/units. The number N is an odd number and is expressed by $N = 2n + 1$. The NMR system will be successful/operational if at least $(n + 1)$ modules/units operate normally. As the voter unit acts in series with the N-module system, the whole system fails whenever the voter unit fails.

For independent modules and voter units, the reliability of the NMR system with voter is given by the following equation [6,9,12]:

$$R_{Nsv} = \left[\sum_{i=0}^{n} \binom{N}{i} R^{N-1}(1-R)^i \right] R_v,$$ (5.4)

$$\binom{N}{i} = \frac{N!}{(N-i)!\,i!},$$ (5.5)

where
  R_v is the voter unit reliability.
  R is the module/unit reliability.
  R_{Nsv} is the NMR system with voter reliability.

It is to be noted that time-dependent reliability analysis of an NMR system can be performed in a manner similar to the TMR system time-dependent reliability analysis presented earlier. Furthermore, information on additional redundancy schemes is available in the study by Nerber [11].

5.5 Internet Failure Examples and Reliability-Related Observations

Each year a large number of Internet-related failures and incidents occur around the globe. For example, in 2001, there were 52,658 Internet-related failures and incidents [2,3,6]. Three examples of Internet failures are presented in the following [6].

- *Example I:* On August 14, 1998, a misconfigured main Internet database server wrongly referred all queries for Internet systems/machines with names ending in *net* to the incorrect secondary database server. Due to this problem, most of the connections to net Internet servers and other end stations operated incorrectly for a number of hours [2,6].
- *Example II:* On April 23, 1997, a misconfigured router of a Virginia service provider injected a wrong map into the global Internet. The Internet providers who accepted this map automatically diverted their traffic to the Virginia provider [6,13]. This resulted in network instability, congestion, and overload of Internet router table memory that shut down many of the main Internet backbones for almost 2 hours [2,6,13].

- *Example III:* On November 8, 1998, a malformed routing control message because of a software-related fault triggered an interoperability problem between many core Internet backbone routers produced by different vendors. This resulted in a widespread loss of network connectivity in addition to an increment in packet loss and latency [2,5]. It took many hours for most of the backbone providers to correct this outage.

A study reported the following Internet reliability-related observations [14]:

- Most interprovider path failures occur from congestion collapse.
- Most of the Internet backbone paths' MTTF and mean time to repair are approximately 25 days or less and 20 minutes or less, respectively.
- In the Internet backbone infrastructure, there is only a small fraction of network paths that disproportionately contribute to long-term outages and backbone unavailability.
- Availability and MTTF of the Internet backbone structure are significantly less than the Public Switched Telephone Network.

5.6 Internet Outage Classifications

Experiences over the years indicate that there are many types of Internet outages. A case study of Internet outages performed over a 1-year period categorized the outages along with their occurrence percentages under the following 12 classifications [14]:

- *Classification I:* Maintenance—16.2%
- *Classification II:* Power outage—16%
- *Classification III:* Fiber cut/circuit/carrier problem—15.3%
- *Classification IV:* Unreachable—12.6%
- *Classification V:* Hardware problem—9%
- *Classification VI:* Interface down—6.2%
- *Classification VII:* Routing problem—6.1%
- *Classification VIII:* Miscellaneous—5.9%
- *Classification IX:* Unknown/undetermined/no problem—5.6%
- *Classification X:* Congestion/sluggish—4.6%
- *Classification XI:* Malicious attacks—1.5%
- *Classification XII:* Software problems—1.3%

5.7 A Method for Automating Fault Detection in Internet Services and Models for Conducting Internet Reliability and Availability Analyses

Experiences, over the years, clearly indicate that many Internet-related services (e.g., search engines and e-commerce) suffer faults, and a quick detection of these faults could be a critical factor to improve system availability. Thus, for this very purpose, an approach known as the *pinpoint method* is considered quite useful. The method combines the low-level monitors' easy deployability with the higher-level monitors' ability for detecting application-level faults [6,15]. In regard to the system under observation and its workload, the pinpoint method is based upon the following assumptions [6,15]:

- The software under consideration is made up of a number of inter-connected modules with properly defined narrow interfaces, which could be software subsystems, objects, or simply physical mode boundaries.
- There are a considerably higher number of basically independent requests from various different users.
- An interaction with the system is short lived, the processing of which can be decomposed as a path or, more clearly, a tree of the names of elements/parts that take part in the servicing of that request.

The pinpoint method is a three-stage process, and each of its stages is described in the following [6,15].

- *Observing the system:* This stage is concerned with capturing the run-time path of each request served/handled by the system and then, from these paths, extracting two specific low-level behaviors that are most likely for reflecting high-level functionality (i.e., interactions of parts/components and path shapes).
- *Learning the patterns in system behavior:* This stage is concerned with constructing a reference model that represents the normal behavior of an application in regard to part/component interactions and path shapes. The model is constructed under the assumption that most of the time, the system functions normally.
- *Detecting anomalies in system behaviors:* This stage is concerned with performing an analysis of the ongoing behaviors of the system as well as detecting anomalies with respect to the reference model.

All in all, additional information on pinpoint method is available in the study by Kiciman and Fox [15].

There are many mathematical models in the published literature that can be used to perform various types of Internet reliability and availability analyses [2,9,16–19]. Two such models are presented in the following.

5.7.1 Mathematical Model I

Mathematical model I is concerned with evaluating the reliability and the availability of an Internet server system when it can be in either an operating or a failed state. Furthermore, the model assumes that all its failures/outages occur independently, the repaired/restored server system is as good as new, and its failure/outage and repair/restoration rates are constant.

The Internet server system state space diagram is shown in Figure 5.3, and the numerals in the box and the circle denote system states.

The following symbols are associated with this mathematical model:

i is the ith server system state shown in Figure 5.3: $i = 0$ (means server system operating normally); $i = 1$ (means server system failed).

$P_i(t)$ is the probability that the server system is in state i at time t for $i = 0, 1$.

λ_s is the server system constant failure/outage rate.

θ_s is the server system constant repair/restoration rate.

With the aid of the Markov method presented in Chapter 4, we get the following differential equations for the diagram shown in Figure 5.3 [6,9]:

$$\frac{dP_0(t)}{dt} + \lambda_s P_0(t) = \theta_s P_1(t), \tag{5.6}$$

$$\frac{dP_1(t)}{dt} + \theta_s P_1(t) = \lambda_s P_0(t). \tag{5.7}$$

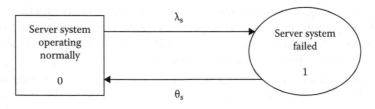

FIGURE 5.3
Internet server system state space diagram.

At time $t = 0$, $P_0(0) = 1$ and $P_1(0) = 0$.

By solving Equations 5.6 and 5.7, we obtain the following equations:

$$P_0(t) = AV_s(t) = \frac{\theta_s}{(\lambda_s + \theta_s)} + \frac{\lambda_s}{(\lambda_s + \theta_s)} e^{-(\lambda_s + \theta_s)t}, \tag{5.8}$$

$$P_1(t) = UA_s(t) = \frac{\lambda_s}{(\lambda_s + \theta_s)} - \frac{\lambda_s}{(\lambda_s + \theta_s)} e^{-(\lambda_s + \theta_s)t}, \tag{5.9}$$

where

$AV_s(t)$ is the Internet server system availability at time t.

$UA_s(t)$ is the Internet server system unavailability at time t.

As time t becomes very large, Equations 5.8 and 5.9 reduce to

$$AV_s = \lim_{t \to \infty} AV_s(t) = \frac{\theta_s}{\lambda_s + \theta_s}, \tag{5.10}$$

$$UA_s = \lim_{t \to \infty} UA_s(t) = \frac{\lambda_s}{\lambda_s + \theta_s}, \tag{5.11}$$

where

AV_s is the Internet server system steady-state availability.

UA_s is the Internet server system steady-state unavailability.

For $\theta_s = 0$, Equation 5.8 becomes

$$R_s(t) = e^{-\lambda_s t}, \tag{5.12}$$

where

$R_s(t)$ is the Internet server system reliability at time t.

Thus, the Internet server system MTTF is expressed by Dhillon [9] as

$$MTTF_s = \int_0^\infty R_s(t)\, dt$$

$$= \int_0^\infty e^{-\lambda_s t}\, dt \tag{5.13}$$

$$= \frac{1}{\lambda_s}.$$

Example 5.2

Assume that the constant failure and repair rates of an Internet server system are 0.002 failures/hour and 0.04 repairs/hour, respectively. Calculate the server system unavailability for a 15-hour mission.

By substituting the given data values into Equation 5.9, we get

$$UA_s(15) = \frac{0.002}{0.002+0.04} - \frac{0.002}{0.002+0.04}e^{-(0.002+0.04)(15)}$$

$$= 0.0222.$$

Thus, the Internet server system unavailability for the specified mission time is 0.0222.

5.7.2 Mathematical Model II

This mathematical model is concerned with evaluating the availability of an Internet working (router) system composed of two independent and identical switches. The model assumes that the switches form a standby-type network and that the system malfunctions when both the switches malfunction. Furthermore, the failure and restoration/repair rates of the switches are constant. The state space diagram of the system is shown in Figure 5.4, and the numerals in the diagram circles and box denote system states.

The following symbols are associated with this model:

i is the ith system state shown in Figure 5.4 for $i = 0$ (system operating normally [i.e., two switches functional: one is operating; the other is on standby]), $i = 1$ (one switch is operating; the other failed), and $i = 2$ (system failed [both switches failed]).

$P_i(t)$ is the probability that the Internetworking (router) system is in state i at time t for $i = 0, 1, 2$.

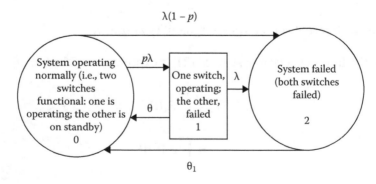

FIGURE 5.4
System state space diagram.

λ is the switch constant failure rate.

θ is the switch constant repair/restoration rate.

θ_1 is the constant restoration/repair rate from system state 2 to state 0.

p is the probability of failure detection and successful switchover from switch failure.

With the aid of the Markov method presented in Chapter 4, we get the following differential equations for the diagram shown in Figure 5.4 [6,9]:

$$\frac{dP_0(t)}{dt} + [p\lambda + (1-p)\lambda]P_0(t) = \theta P_1(t) + \theta_1 P_2(t), \tag{5.14}$$

$$\frac{dP_1(t)}{dt} + (\lambda + \theta)P_1(t) = p\lambda P_0(t), \tag{5.15}$$

$$\frac{dP_2(t)}{dt} + \theta_1 P_2(t) = \lambda P_1(t) + \lambda(1-p)P_0(t). \tag{5.16}$$

At time $t = 0$, $P_0(0) = 1$, $P_1(0) = 0$, and $P_2(0) = 0$.

The following steady-state probability equations are obtained by setting derivatives equal to zero in Equations 5.14 through 5.16 and using the relationship $\sum_{i=0}^{2} P_i = 1$:

$$P_0 = \theta_1(\theta + \lambda)/A, \tag{5.17}$$

where

$$A = \theta_1(\theta + p\lambda + \lambda) + (1-p)\lambda(\theta + \lambda) + p\lambda^2,$$
$$P_1 = p\lambda\theta_1/A, \tag{5.18}$$

$$P_2 = [p\lambda^2 + (1-p)\lambda(\theta + \lambda)]/A, \tag{5.19}$$

where

P_i is the steady-state probability that the Internetworking (router) system is in state i for $i = 0, 1, 2$.

The Internetworking (router) system steady-state availability is expressed by

$$AV_{is} = P_0 + P_1,$$

$$AV_{is} = [\theta_1(\theta + \lambda) + p\lambda\theta_1]/A, \tag{5.20}$$

where
AV_{is} is the Internetworking (router) system steady-state availability.

5.8 Robot Reliability-Related Survey Results and Effectiveness Dictating Factors

Jones and Dawson [20] reported the findings/results of a robot reliability-related survey study of 37 robots of four different designs being used in three different companies X, Y, and Z, covering 21,932 robot production hours. The three companies X, Y, and Z reported 47, 306, and 155 cases of robot reliability-associated problems, respectively, of which the corresponding 27, 35, and 1 cases did not contribute to any downtime at all. More clearly, for companies X, Y, and Z, robot downtime as a percentage of production time was 1.8%, 13.6%, and 5.1%, respectively.

Approximate figures for mean time to robot-related problems (MTTRPs) and mean time to robot failures (MTTRFs) for companies X, Y, and Z are presented in Table 5.1 [20].

It is to be noted from Table 5.1 that among these three companies, there is a quite wide variation in MTTRP and MTTRF. More specifically, highest and lowest MTTRPs and MTTRFs are 221 and 15 hours and 2596 and 40 hours, respectively.

There are a large number of factors that dictate the effectiveness of robots. Some of these factors are as follows [21–23]:

- The percentage of time the robot operates normally
- Robot mean time between failures

TABLE 5.1

Approximate MTTRPs and MTTRFs
for Companies X, Y, and Z

Company	MTTRP (Hours)	MTTRF (Hours)
X	221	2596
Y	30	284
Z	15	40

- The relative performance of the robot under extreme conditions
- Robot mean time to repair
- Quality and availability of personnel needed to keep the robot in operating state
- Quality and availability of robot repair facilities and equipment
- The percentage of time the robot is available for operation
- Rate of the availability of the required spare parts/components

5.9 Categories of Robot Failures and Their Causes and Corrective Measures

There are various types of robot failures, and they can be grouped under the four categories shown in Figure 5.5 [21,23–25].

Category I: Systematic hardware faults are failures which occur because of the existence of unrevealed mechanisms in the root system design. Some of the reasons for the occurrence of systematic faults are as follows:

- Peculiar wrist orientations
- Unusual joint-to-straight-line mode transition
- Failure to make the appropriate environment-related provisions in the initial design

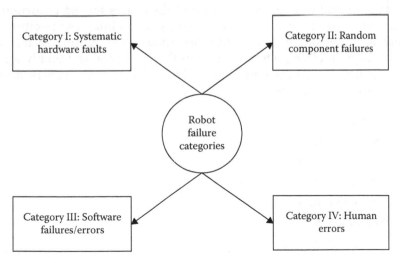

FIGURE 5.5
Categories of robot failures.

Some of the methods that can be used to reduce the occurrence of robot systematic faults or failures are the inclusion of sensors in the system for detecting the loss of pneumatic pressure, line voltage, or hydraulic pressure and the use of sensors for detecting excessiveness of force, speed, server errors, temperature, and acceleration.

Several methods useful for reducing systematic faults or failures are described by Dhillon [5,24].

Category II: Random component failures are failures that occur unpredictably during the component's useful life. Some of the reasons or the causes for the occurrence of such failures are undetectable defects, low safety factors, unavoidable failures, and unexplainable causes/reasons.

Some of the methods presented in Chapter 4 and in the study by Dhillon [9] can be used to reduce such failures' occurrence.

Category III: Software failures/errors are associated with software concerned with robots. In robots, software failures/errors/faults can occur in the embedded software or the controlling software and the application software. As per one study reported by Dhillon [5], over 60% of software errors are made during the requirement and the design phase as opposed to less than 40% during the coding phase.

Although redundancy is expensive, it is probably the best solution to protect against the occurrence of software failures or errors. Also, the use of approaches such as failure modes and effect analysis, fault tree analysis, and testing can be quite useful for reducing software failures or errors. Furthermore, there are many software reliability models that can be used to evaluate reliability when the software is put into operational use [5,9,4,25].

Category IV: Human errors are due to personnel who maintain, operate, test, manufacture, and design robots. Some of the causes for the occurrence of human errors are poor equipment design, poor training of operating and maintenance personnel, task complexity, inadequate lighting in the work area, improper tools, high temperature in the work area, and poorly written operating and maintenance procedures. Thus, human errors may be divided into classifications such as follows:

- Design errors
- Operating errors
- Maintenance errors
- Inspection errors
- Assembly errors
- Installation errors

Some of the methods that can be used to reduce the occurrence of human errors are error cause removal program, human–machine system analysis, fault tree analysis, and probability tree analysis. The first two methods are described in the study by Dhillon [26], and the remaining two methods are described in Chapter 4.

5.10 Robot Reliability Measures and Analysis Methods

There are various types of measures and methods used in performing robot reliability analysis. Both of these items are presented in the following separately.

5.10.1 Robot Reliability Measures

The three commonly used robot reliability measures are presented in the following sections.

5.10.1.1 Mean Time to Robot-Related Problems

The mean productive robot time before the occurrence of a robot-related problem is expressed by

$$MTTRP = \frac{RPT - DDTRP}{NRP},\tag{5.21}$$

where
$MTTRP$ is the MTTRP.
RPT is the robot production time expressed in hours.
$DDTRP$ is the downtime due to robot-related problems expressed in hours.
NRP is the number of robot-related problems.

Example 5.3

Assume that at an industrial facility, the annual robot production hours and downtime due to robot-related problems are 10,000 and 250 hours, respectively. There were 20 robot-related problems during the 1-year period. Calculate the MTTRP.

By inserting the specified data values into Equation 5.21, we obtain

$$MTTRP = \frac{10,000 - 250}{20}$$

$$= 487.5 \text{ hours.}$$

Thus, the MTTRP is 487.5 hours.

5.10.1.2 Mean Time to Robot Failure

The MTTRF can be obtained by using any of the following three equations [6,9]:

$$MTRF = \int_0^\infty R_r(t)\,dt, \tag{5.22}$$

$$MTRF = \lim_{s \to 0} R_r(s), \tag{5.23}$$

$$MTRF = \frac{RPT - DDTRF}{NRF}, \tag{5.24}$$

where
$MTRF$ is the MTTRF.
$R_r(t)$ is the robot reliability at time t.
$R_r(s)$ is the Laplace transform of the robot reliability at time t, $R_r(t)$.
NRF is the number of robot failures.
$DDTRF$ is the downtime due to robot failures expressed in hours.

Example 5.4

Assume that the constant failure rate λ of an industrial robot is 0.0001 failures/hour and its reliability is expressed by

$$R(t) = e^{-\lambda t},$$
$$= e^{-(0.0001)t}, \tag{5.25}$$

where
$R(t)$ is the industrial robot reliability at time t.

Calculate the MTTRF by using Equations 5.22 and 5.23 and comment on the result.
By substituting Equation 5.25 into Equation 5.22, we obtain

$$MTRF = \int_0^\infty e^{-(0.0001)t}\,dt$$

$$= \frac{1}{0.0001}$$

$$= 10,000 \text{ hours.}$$

By taking the Laplace transform of Equation 5.25, we get

$$R(s) = \frac{1}{(s+0.0001)} \cdot$$
(5.26)

By inserting Equation 5.26 into Equation 5.23, we obtain

$$MTRF = \lim_{s \to 0} \frac{1}{(s+0.0001)}$$

$$= \frac{1}{0.0001}$$

$$= 10,000 \text{ hours.}$$

In both cases, the result (i.e., *MTRF* = 10,000 hours) is the same. It proves that Equations 5.22 and 5.23 yield the same result.

5.10.1.3 Robot Reliability

Robot reliability may simply be described as the probability that a robot will carry out its specified function satisfactorily for the stated time interval when used according to the designed conditions. The general formula for obtaining time-dependent robot reliability is as follows [9,21]:

$$R_r(t) = \exp\left[-\int_0^t \lambda_r(t)\,dt\right],$$
(5.27)

where
$R_r(t)$ is the robot reliability at time t.
$\lambda_r(t)$ is the time-dependent failure rate (hazard rate) of the robot.

This means that Equation 5.27 can be used to obtain reliability function of a robot for any failure time's probability distribution (e.g., exponential, Weibull, or gamma).

Example 5.5

Assume that the times to failure of a robot follow exponential distribution; thus, its hazard rate is constant and is 0.0005 failures/hour. Calculate the robot reliability for a 10-hour mission.

By inserting the robot specified constant hazard rate data value into Equation 5.27, we get

$$R_r(t) = \exp\left[-\int_0^t (0.0005)\,dt\right]$$

$$= e^{-(0.0005)t}.$$

(5.28)

Substituting the specified robot mission time data value into Equation 5.28 yields

$$R_r(10) = e^{-(0.0005)(10)}$$

$$= 0.9950.$$

Thus, the robot reliability for the specified mission time is 0.9950.

5.10.2 Robot Reliability Analysis Methods

There are many methods used to perform various types of reliability analysis in the area of reliability engineering. Some of these methods can be used effectively to perform robot reliability analysis. Four of these methods are described in the following sections.

5.10.2.1 Fault Tree Analysis

The fault tree analysis was developed at the Bell Telephone Laboratories in the 1960s and is widely used in the industry to evaluate the reliability of engineering systems during their design and development phase, in particular the ones used in nuclear power generation. A fault tree may simply be described as a logical representation of the relationship of basic or primary fault events that lead to the occurrence of a stated undesired event known as the *top event*.

Additional information on this method is available in the study by Dhillon [9] and the *Fault Tree Handbook* [27] and in Chapter 4.

5.10.2.2 Failure Modes and Effect Analysis

The failure modes and effect analysis is considered as an effective tool for performing analysis of each failure mode in the system/equipment to determine the effects of such failure modes on the entire system/equipment. The method was developed in the early 1950s by the US DOD [9].

FMEA is composed of following six steps [9,28,29]:

- Step 1: Define system/equipment and its associated requirements.
- Step 2: Develop appropriate ground rules.

- Step 3: Describe the system/equipment and its associated functional blocks.
- Step 4: Highlight possible failure modes and their effects.
- Step 5: Develop a list of critical items.
- Step 6: Document the analysis.

Additional information on this method is available in a book by Dhillon [9], a paper by Coutinho [28], the specification MIL-F-18372 [29], and Chapter 4.

5.10.2.3 Parts Count Method

The parts count method is usually used during bid proposal and early design phases for estimating system/equipment failure rate. The method requires information on items such as part quality levels, system/product/equipment use, and generic parts' quantities and types.

Additional information on this method is available in the study by Dhillon [9] and the report *Reliability Prediction of Electronic Equipment* [30].

5.10.2.4 Markov Method

The Markov method can be used in more cases than any other reliability analysis method. The method is concerned with modeling systems with constant failure and repair rates.

Additional information on the Markov method is available in books by Dhillon [9] and Shooman [31] and in Chapter 4.

PROBLEMS

1. List and describe the main sources of computer failures.
2. Describe the computer-related fault classifications.
3. Describe the following terms:
   a. *Fault masking*
   b. *Triple modular redundancy*
   c. *N-modular redundancy*
4. Assume that the constant failure rate of a module/unit of a TMR with voter is 0.0003 failures/hour and the constant failure rate of the voter unit is 0.0001 failures/hour. Calculate the system mean time to failure and reliability for a 200-hour mission.
5. Discuss at least three examples of Internet failures.
6. Describe the pinpoint method.
7. Prove Equations 5.17 through 5.19 by using Equations 5.14 through 5.16.

8. Discuss the following two categories of robot failures:
   a. Systematic hardware faults
   b. Software failures/errors
9. Assume that at an industrial facility, the annual robot production and downtime hours due to robot-related problems are 8000 and 200 hours, respectively. There were 10 robot-related problems during the 1-year period. Calculate the MTTRPs.
10. Describe at least three methods that can be used to perform robot reliability analysis.

References

1. Hafner, K., Lyon, M., *Where Wizards Stay Up Late: The Origin of the Internet*, Simon and Schuster, New York, 1996.
2. Dhillon, B.S., *Applied Reliability and Quality: Fundamentals, Methods, and Procedures*, Springer, Inc., London, 2007.
3. Goseva-Popstojanova, K., Mazidar, S., Singh, A.D., Empirical study of session-based workload and reliability for web servers, *Proceedings of the 15th International Symposium on Software Reliability Engineering*, 2004, pp. 403–414.
4. Goldberg, J., *A Survey of the Design and Analysis of Fault-Tolerant Computers, in Reliability and Fault Tree Analysis*, edited by R.E. Barlow, J.B. Fussell, and N.D. Singpurwalla, Society for Industrial and Applied Mathematics, Philadelphia, PA, 1975, pp. 667–685.
5. Dhillon, B.S., *Reliability in Computer System Design*, Ablex Publishing, Norwood, NJ, 1987.
6. Dhillon, B.S., *Computer System Reliability: Safety and Usability*, CRC Press, Boca Raton, FL, 2013.
7. Yourdon, E., The causes of computer failures: Part III, *Modern Data*, Vol. 5, 1972, pp. 50–56.
8. Yourdon, E., The causes of computer failures: Part II, *Modern Data*, Vol. 5, 1972, pp. 36–40.
9. Dhillon, B.S., *Design Reliability: Fundamentals and Applications*, CRC Press, Boca Raton, FL, 1999.
10. Avizienis, A., Fault-tolerant computing: Progress, problems, and prospectus, *Proceedings of the International Federation for Information Processing Congress*, 1977, pp. 405–420.
11. Nerber, P.O., Power-off time impact on reliability estimates, *IEEE International Convention Record*, Part 10, March 1965, pp. 1–8.
12. Shooman, M.L., *Reliability of Computer Systems and Networks: Fault Tolerance, Analysis, and Design*, Wiley, New York, 2002.
13. Barrett, R., Haar, S., Whitestone, R., Routing snafu causes Internet outage, *Interactive Week*, April 25, 1997, p. 9.

14. Lapovitz, C., Ahuja, A., Jahamian, F., Experimental study of the Internet stability and wide-area backbone failures, *Proceedings of the 29th Annual International Symposium on Fault-Tolerant Computing*, 1999, pp. 278–285.
15. Kiciman, E., Fox, A., Detecting application-level failures in component-based Internet services, *IEEE Transactions on Neural Networks*, Vol. 16, No. 5, 2005, pp. 1027–1041.
16. Aida, M., Abe, T., Stochastic model of Internet access patterns, *IEICE Transactions on Communications*, Vol. E84-B(8), 2001, pp. 2142–2150.
17. Hecht, M., Reliability/Availability modeling and prediction of e-commerce and other Internet information systems, *Proceedings of the Annual Reliability and Maintainability Symposium*, 2001, pp. 176–182.
18. Imaizumi, M., Kimura, M., Yasui, K., Optimal monitoring policy for server system with illegal access, *Proceedings of the ISSAT International Conference on Reliability and Quality in Design*, 2005, pp. 155–159.
19. Chan, C.K., Tortorella, M., Spare-inventory sizing for end-to-end service availability, *Proceedings of the Annual Reliability and Maintainability Symposium*, 2001, pp. 98–102.
20. Jones, R., Dawson, S., People and robots: Their safety and reliability, *Proceedings of the 7th British Robot Association Annual Conference*, 1984, pp. 243–258.
21. Dhillon, B.S., *Robot Reliability and Safety*, Springer-Verlag, New York, 1991.
22. Yong, J.F., *Robotics*, Butterworth, London, 1973.
23. Dhillon, B.S., *Robot System Reliability and Safety: A Modern Approach*, CRC Press, Boca Raton, FL, 2015.
24. Dhillon, B.S., *Reliability Engineering in Systems Design and Operation*, Van Nostrand Reinhold Company, New York, 1983.
25. Herrmann, D.S., *Software Safety and Reliability*, IEEE Computer Society Press, Los Alamitos, CA, 1999.
26. Dhillon, B.S., *Human Reliability, Error, and Human Factors in Power Generation*, Springer, London, 2014.
27. Report No. NUREG-0492, *Fault Tree Handbook*, US Nuclear Regulatory Commission, Washington, DC, 1981.
28. Coutinho, J.S., Failure effect analysis, *Transactions of the New York Academy of Sciences*, Vol. 26, Series II, 1963–1964, pp. 564–584.
29. MIL-F-18372 (Aer), *General Specification for Design, Installation, and Test of Aircraft Flight Control Systems*, Bureau of Naval Weapons, Department of the Navy, Washington, DC, Para, 3.5.2.3.
30. MIL-HDBK-217, *Reliability Prediction of Electronic Equipment*, Department of Defense, Washington, DC.
31. Shooman, M.L., *Probabilistic Reliability: An Engineering Approach*, McGraw-Hill Book Company, New York, 1968.

6

Transportation System Failures and Human Error in Transportation Systems

6.1 Introduction

Each year, a vast sum of money is spent around the globe to develop, manufacture, and operate transportation systems such as motor vehicles, aircraft, ships, and trains. Throughout the world, transportation systems such as these carry billions of passengers and billions of tons of goods from one point to another annually. For example, according to the International Air Transport Association, each year, the world's airlines carry over 1.6 billion passengers for business and leisure travels, and around 40% of global trade of goods is carried by air [1].

Needless to say, transportation system failures and human error in transportation systems have become an important issue, because they can, directly or indirectly, impact the global economy and the environment, in addition to transportation safety and reliability. For example, in regard to road transportation system safety only, each year, around 0.8 million road accident fatalities and 20–30 million injuries occur around the globe [2,3]. It is to be noted that human error is considered to be an important factor in the occurrence of such events.

This chapter presents various important aspects of transportation system failures and human error in transportation systems.

6.2 Defects in Vehicle Parts and Categories of Vehicle Failures

A motor vehicle is composed of many parts and subsystems such as brakes, steering system, rim, engine, clutch, and transmission. The malfunctioning of parts and subsystems such as these can lead to motor vehicle failure.

Defects in brake, steering system, and rim are discussed in the following, separately [4,5].

- *Brake defects:* In normal driving environments, the malfunctioning of parts in the braking system of the motor vehicle is likely to occur only when the parts become severely worn, defective, or degraded. Brake defects may be grouped under four classifications as shown in Figure 6.1.

  Air brake system defects include slow pressure buildup in the reservoir, low or no brake force, and slow brake response or release. Some of the defects belonging to the disk brake system defect classification are low or no brake force, excessive wear of the pad, and excessive brake pedal travel.

  The common disk and drum brake system defects include items such as excessive pedal force, brake pedal vibrations, brake fade, and soft pedal. Finally, some of the defects belonging to the drum brake system defect classification are noise generation during braking, brake jam, increasing heat in the brakes while driving the vehicle, brake imbalance, brake pedal touching floor, and low braking performance and hard pedal.

- *Steering system defects:* These defects can result in severe motor vehicle accidents. There are many causes for the occurrence of steering system defects. Some of these causes are poor maintenance, faulty design, faulty changes made to the steering system, faulty manufacturing, and inadequate inspection.

- *Rim defects:* These types of defects are as important as defects in any other important part of a motor vehicle, because they can lead to

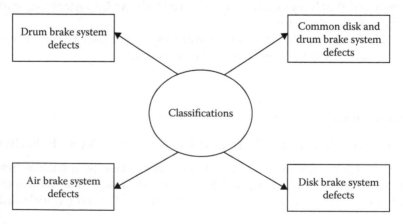

FIGURE 6.1
Classifications of brake defects.

serious accidents. As per the study by Anderson [4], 1 in about 1300–2200 truck tire failures leads to an accident, and as per the findings of the US Bureau of Motor Carrier Safety, approximately 7–13% of trailers and tractors had at least one defective tire.

Some of the causes of the rim defects are faulty manufacturing operations, abusive operation, and poor design.

Failures of a vehicle carrying passengers can be grouped under categories A, B, C, and D and are as follows [5,6]:

- *Category A:* In this case the vehicle stops or is required to stop and is pushed/towed by an adjacent vehicle to the close by station. At this point, individuals in both the affected vehicles egress, and the failed vehicle is pushed/towed for maintenance.
- *Category B:* In this case the vehicle stops and it cannot be towed or pushed by the adjacent vehicle, and it must wait for the rescue vehicle.
- *Category C:* In this case the vehicle is required to reduce speed and is allowed to continue to the closest station, where its passengers must egress, and then, it is taken for maintenance.
- *Category D:* In this case the vehicle is allowed to continue to the nearest station, where its passengers must egress, and then, it is taken for maintenance.

6.3 Rail Weld Failures and Defects

In railway systems, the construction of continuous welded rails (CWRs) is indispensable to reduce the vibration and the noise, improve the ride quality, and reduce the track maintenance cost. Over the years, due to rail weld failures, many railway accidents have occurred. Thus, it is important to have highly reliable welds in order to eradicate the occurrence of weld-related failures in service and to extend the service life of CWR.

Data collected over the years clearly indicate that most rail weld-related failures are initiated from weld discontinuities, and fusion welding tends to easily cause such discontinuities [6]. Therefore, fusion welding methods such as aluminothermic welding and enclosed arc welding are less reliable than pressure welding methods such as gas pressure welding and flash welding [6,7].

Thus, to eliminate the rail weld failures' occurrence, it is important to perform reliable welding by using proper welding processes, welding conditions, inspection approaches, and well-trained welding technicians.

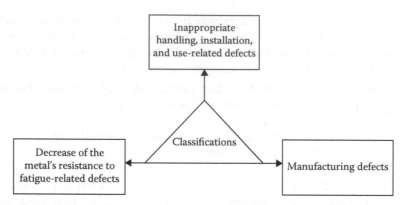

FIGURE 6.2
Classifications of defects in steel rail.

Although the hardness basically controls the wear resistance of rails, the wear resistance is also dependent on the stresses that the rails are subjected to. These stresses include contact stresses, bending stresses, thermal stresses, and residual stresses, and they control the development of defects in rails that can eventually lead to failure [8,9].

The contact stresses originate from the wheel load, the traction, and the braking and the steering actions. The bending stresses act either laterally or vertically, and the vertical ones are mainly tensile in the rail base and compressive in the railhead. The thermal stresses originate from welding processes during the connection of rail sections for creating a continuously welded rail, whereas the residual stresses originate from manufacturing processes.

Defects in steel rails may be grouped under three classifications as shown in Figure 6.2 [9].

The decrease of the resistance of the metal to fatigue-related defects includes the most common rail defects such as squats and head checks. The manufacturing defects originate from the rail manufacturing process. Finally, the inappropriate handling and installation and the use-related defects originate from out-of-specification installation of rails, wheel burns, and unexpected scratches.

6.4 Classifications of Road and Rail Tanker Failure Modes and Causes of Failures and Factors Influencing the Nature of Failure Consequences

Road and rail tankers are used for carrying liquefied gas and other hazardous liquids from one point to another point. Over the years, the failure of such tankers has led to serious consequences. The main locations of the tanker failures are pumps, inspection covers, valves, shells, connections to a container, and branches, including instrument connections.

The failure modes of road and rail tankers may be grouped under three classifications as shown in Figure 6.3 [10].

The main causes for metallurgical failures are as follows:

- Erosion
- Corrosion (internal and external)
- Fatigue
- Vessel used for purposes not covered by specification
- Vessel designed/constructed to an inadequate specification
- Failure to satisfy specified construction codes
- Use of wrong or inadequate materials of construction
- Embrittlement by chemical action

The main causes for failures due to mechanical causes other than overpressure include collision with another vehicle, damage by an external explosion, general tear and wear, collapse of a structure onto it, modifications in violation of original specifications, and collision with a fixed object such as a bridge. Finally, the main causes for failures due to excess internal pressure are tanker contents having higher vapor pressure than designed for, abnormal meteorological conditions, internal chemical reaction such as decomposition and polymerization, flame impingement, and hydraulic rupture consequent upon overfilling.

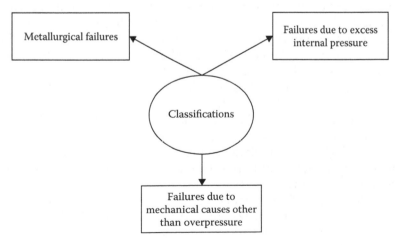

FIGURE 6.3
Classifications of road and rail tanker failure modes.

There are various consequences associated with road and rail tanker failures involving the loss of containment. The nature of such consequences is influenced by the following five principal factors [5,10]:

- The location and the size of any leak which develops
- The physical state of the contents
- The chemical nature of the contents
- The mechanism of dispersion
- The nature of the surroundings

Additional information on the five influencing factors previously mentioned is available in the study by Marshall [10].

6.5 Mechanical Failure-Related Aviation Accidents and Their Examples

Over the years, many aviation accidents have occurred due to mechanical failures and mechanical-related pilot errors. More clearly, the mechanical-related pilot errors are the ones in which pilot errors were the actual cause but brought about by some kind of mechanical failures.

A worldwide study of 1300 fatal accidents during the period of 1950–2008, involving commercial aircraft (i.e., excluding helicopters and aircraft with 10 or fewer individuals on board), reported a total of 134 accidents due to mechanical failure and 25 accidents due to mechanical-related pilot error [5]. It is to be noted here that these two types of accidents (i.e., 134 and 25) are out of those accidents whose causes were clearly identifiable.

Some of the examples of the aviation accidents that occurred due to mechanical failure are United Airlines Flight 585 accident, US Air Flight 427 accident, British International Helicopters Chinook Accident, Turkish Airlines Flight 981 accident, Los Angeles Airways Flight 841 accident, and United Airlines Flight 859 accident.

The United Airlines Flight 585 accident occurred on March 3, 1991, and is concerned with the United Airlines Flight 585 (aircraft type: Boeing 737-291), a scheduled flight from Stapleton International Airport, Denver, Colorado, to Colorado Springs, Colorado [11]. The flight crashed due to rudder device failure and resulted in 25 fatalities.

The US Air Flight 427 accident occurred on September 8, 1994, and is concerned with the US Air Flight 427 (aircraft type: Boeing 737-387), a scheduled flight from O'Hare Airport, Chicago, Illinois, to West Palm Beach, Florida, via Pittsburgh, Pennsylvania [12]. The flight crashed due to rudder device failure and resulted in 132 fatalities.

The British International Helicopters Chinook accident occurred on November 6, 1986, and is concerned with a Boeing 234LR Chinook helicopter operated by a company named British International Helicopters [13]. The helicopter on approach to land at Sumburgh Airport, Shetland Islands, United Kingdom, crashed into the sea and sank due to the failure of a modified level ring gear in the forward transmission. The accident caused 18 fatalities and 84 injuries.

The Turkish Airlines Flight 981 accident occurred on March 3, 1974, and is concerned with the Turkish Airlines Flight 981 (aircraft type: McDonnell Douglas DC-10-10), a scheduled flight from Istanbul, Turkey, to Heathrow Airport, London, United Kingdom, via Paris, France [14]. The flight crashed due to cargo hatch malfunction and control cable failures and resulted in 346 fatalities.

The Los Angeles Airways Flight 841 accident occurred on May 22, 1968, and is concerned with the Los Angeles Airways Flight 841 (aircraft type: Sikorsky S-611 helicopter), a scheduled flight from Disneyland heliport, Anaheim, California, to Los Angeles International Airport, Los Angeles, California [15]. The flight crashed due to a mechanical malfunction in the blade rotor system and resulted in 23 fatalities.

Finally, the United Airlines Flight 859 accident occurred on July 11, 1961, and is concerned with the United Airlines Flight 859 (aircraft type: Douglas DC-8-20), a scheduled flight from Omaha, Nebraska, to Stapleton International Airport, Denver, Colorado [16]. The flight crashed during landing at the Stapleton International Airport because the aircraft suffered a hydraulic malfunction while en route and resulted in 18 fatalities and 84 injuries.

6.6 Ship Failures and Their Common Causes

The shipping industrial sector is composed of many types of ships including bulk cargo ships, tankers, container ships, and carriers. Ships such as these contain various types of systems, equipment, and parts that can occasionally fail. Some of the examples of such systems, equipment, and part failures are shown in Figure 6.4 [5].

It is to be noted that the consequences of the failures of the items shown in Figure 6.4 can vary quite considerably. Nonetheless, some of the common causes for the occurrence of ship failures are as follows [5]:

- Poor quality assurance
- Manufacturing defects
- Fatigue

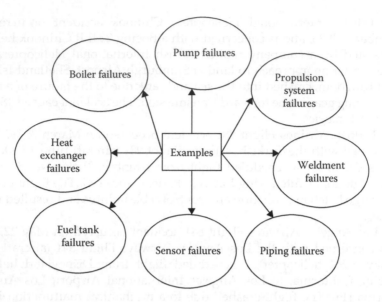

FIGURE 6.4
Examples of ship systems, equipment, and part failures.

- Unforeseen operating conditions
- Corrosion
- Welding defects
- Improper maintenance

6.7 Railway System Human Error-Related Facts and Figures and Typical Human Error Occurrence Areas in Railway Operation

Some of the railway system's direct or indirect human error-related facts and figures are presented in the following:

- As per the study by Reinach and Viale [17], in 2004, approximately 53% of the railway switching yard accidents (i.e., excluding highway-rail crossing train accidents) in the United States were due to human factors-associated causes.
- During the period from 1970 to 1998, in Norway, approximately 62% of 13 railway accidents that resulted in injuries or fatalities were due to human errors [18].

- As per the study by *The Nation Newspaper* [19], in 2005, a subway train crash at Thailand Cultural Center Station, Bangkok, due to a human error injured around 200 people.
- During the period from 1900 to 1997, in the United Kingdom, approximately 70% of the 141 accidents on four British Railway main lines were due to human errors [18,20].
- As per Railway Board of the Ministry of Railways [21], in India, each year, over 400 railway accidents occur, and approximately 66% of these accidents, directly or indirectly, are the result of human error-related causes.
- In the United Kingdom, in 1999, 31 persons died and 227 people were hospitalized in train accidents due to human error [22].

In railway operation, there are many areas in which human errors can occur. Three typical such areas are as follows [18]:

- Signaling or dispatching
- Signal passing
- Train speed

In the case of signaling or dispatching, over the years, many railway accidents have occurred because of dispatchers or signalperson errors. Nowadays, with the use of modern technology/devices, human errors in the area of signaling or dispatching have been reduced significantly [18].

In the case of signal passing, over the years, many railway accidents have occurred. Trains passing a signal displaying a stop is very dangerous, as it can lead to an immediate conflict with another train or trains. Often, this event/situation is referred to as signal passed at danger (SPAD). Each year, around the globe, many SPAD-related incidents occur. For example, for the British Railway System alone, there were 653 such incidents for the period of 1996–1997 [18].

Some of the main causes for the occurrence of SPAD events are as follows [18,23]:

- Overspeeding in regard to braking performance and warning signal distance
- Misjudging the brakes' effectiveness under certain situations, such as bad weather
- Oversight or disregard of a signal
- Failure to see signal due to poor visibility
- The driver falling asleep or being unconscious
- Misjudging of which signal is applicable to the train in question
- Misunderstanding of signaling aspect

Finally, in the case of train speed, over the years, many accidents have occurred due to the failure of train drivers to reduce the train speed as specified for the route in question. The likelihood of overspeeding and its associated consequences depend on a number of factors including the circumstances surrounding it and the type of speed restrictions.

There are basically three types of speed restrictions that require proper driver response from his or her perspective:

1. *Permanent speed restrictions:* These speed restrictions are imposed because of track curves or some existing infrastructure-associated conditions on a certain portion of a track in question.

2. *Temporary or emergency speed restrictions:* These speed restrictions are imposed because of temporary track shortcomings such as stability-related problems and frost heave or maintenance work.

3. *Conditional speed restrictions:* These speed restrictions are imposed because of train route setting at a particular junction or station and the signaling aspect displayed in that regard.

6.8 Aviation System Human Error-Related Facts and Figures and Types of Pilot–Controller Communication-Related Errors

Some of the aviation system human error-associated facts and figures are as follows:

- A study conducted by Boeing Commercial Airplane Group reported that in the occurrence of over 73% of aircraft accidents worldwide, the failure of the cockpit crew has been a contributory factor [24,25].

- As per the study by *Science Daily* [26], during the period 1990–1996, pilot error was responsible for approximately 34% of major airline crashes.

- During the period of 1983–1996, there were 371 major airline crashes; 29,798 general aviation crashes; and 1735 commuter/air taxi crashes [26]. A study of these crashes clearly indicated that pilot error was a probable cause for approximately 38% of major airline crashes,

85% of general aviation crashes, and 74% of commuter/air taxi crashes [26].

- As per the studies by Dhillon [23] and *Science Daily* [26], a study reported that in the United States, crashes due to pilot error in major airlines decreased from 43% for the period of 1983–1989 to 34% for the period of 1990–1996.

- A study reported that around 45% of major airline crashes occurring at airports are caused by pilot error as opposed to about 28% of those occurring elsewhere [26,27].

- As per the study by Helmreich [28], a study reported that since the introduction of highly reliable turbojet aircraft in the late 1950s, over 71% of airline accidents involved human error to some degree.

Communication between pilots and air traffic controllers is subject to various types of errors. A study of 386 reports submitted to the Aviation Safety Reporting System from July 1991 to May 1996 indicates that pilot–controller communication-related errors may be categorized under the following four types [29]:

- *Type 1: No pilot readback errors:* As per the study, this type of error accounted for 25% of pilot–controller communication-related errors, and the most common reason for the occurrence of these errors was the pilot expectation [29].

- *Type 2: Readback/hearback errors:* As per the study, this type of error accounted for 47% of pilot–controller communication-related errors and the most common contributing factor for the occurrence of these errors was similar call signs [29].

- *Type 3: Hear back error type 2:* These types of errors are those errors in which a pilot correctly repeats the issued clearance, but the controller overlooks the issued clearance that, in fact, was not the clearance that the controller intended to issue. It is to be noted that this type of error includes events where the pilot made a problematic intent/ statement that the controller should have noticed instantly. As per the study, this type of error (i.e., type 2: hearback error) accounted for 18% of pilot–controller communication-related errors [29].

- *Type 4: Miscellaneous errors:* These types of errors are errors that cannot be categorized under any of the three types previously mentioned. According to the finding of the study, miscellaneous errors accounted for 10% of pilot–controller communication-related errors [29].

6.9 Organization-Related Factors in Commercial Aviation Accidents with Respect to Pilot Error and Recommendations for Reducing Pilot–Controller Communication Errors

Over the years, many studies have been conducted to highlight organization-related factors in the occurrence of commercial aviation accidents with respect to pilot error. One of these studies, for the period of 1990–2000, analyzed the commercial aviation accident data of the National Transportation Safety Board. The findings of this study revealed that during this period, 60 of the 122 accidents were, directly or indirectly, attributable to pilot error due to 70 organization-related factors/causes [30]. These factors/causes are grouped under the following 10 categories [30]:

- *Faulty documentation:* This category includes items such as wrong signoffs, checklists, and record keeping that affect flight operations.
- *Insufficient or untimely information sharing:* This category includes items such as weather reports, logbooks, and updates on the part of the organization.
- *Poor procedures or directives:* This category includes items such as ill-defined or conflicting policies and formal oversight of operation.
- *Poor standards/requirements:* This category includes items such as adherence to policy and clearly defined organizational objectives.
- *Company/management-induced pressures:* This category includes items such as threats to pilot job status and/or pay.
- *Poor supervision of operations at management level:* This category includes items such as failure to provide appropriate guidance, leadership to flight operations, and oversight.
- *Poor facilities:* This category includes items such as failure to provide satisfactory controls, lighting, clearance, etc., for flight operations.
- *Inadequate initial, upgrade, or emergency training/transaction:* This category includes items such as opportunities for pilot training not implemented or made available to appropriate pilots.
- *Poor surveillance of operations:* This category includes items such as organizational climate issues, chain of command, and quality assurance and trend information.
- *Poor substantiation process:* This category includes items such as well-defined and verified process, accountability, standards of operation, regulation, and reporting/recording process.

Some of the recommendations considered useful to reduce communication-related errors between pilots and controllers are as follows [29]:

- Encourage air traffic controllers to keep all involved instructions short with a maximum of four instructions per transmission.
- Encourage all involved controllers to speak slowly and distinctly.
- In the event of having similar call signs on the frequency, encourage all involved aircraft pilots to say their call sign after and before each and every readback.
- Encourage all involved air traffic controllers to treat all read backs as they would treat any other incoming information.
- Encourage all involved aircraft pilots to respond to all types of controller instructions with complete readback of all important components.
- In the event of having similar call signs on the frequency, encourage all involved traffic controllers to continue announcing this fact.
- Encourage all involved controllers to avoid issuing strings of instructions to different aircraft.

6.10 Shipping System Human Error-Related Facts and Figures

Some of the facts and figures of the shipping systems directly or indirectly related to human error are as follows:

- Each year, according to the findings of the United Kingdom Protection and Indemnity (UKP&I) club, the occurrence of human errors costs the marine industrial sector $541 million [31].
- A study of 6091 accident claims greater than $100,000 concerning all classes of commercial ships over a period of 15 years, conducted by the UKP&I club, reported that approximately 62% of the claims were directly or indirectly attributable to human error [31–33].
- As per the studies by Dhillon [27] and *The Scandinavian Shipping Gazette* [34], in 2004, Bow Mariner, a chemical/product tanker, sunk because of an onboard explosion due to a human error and resulted in 18 deaths of crew members.
- As per the studies by the Transportation Safety Board of Canada [35] and Rothblun [36], the occurrence of human errors contributed 84–88% of tanker-related accidents.
- Human error is a factor in the occurrence of 89–96% of ship collisions [36,37].

- Over 80% of marine-associated accidents are directly or indirectly caused or influenced by human and organization factors [38,39].
- As per the studies by Rothblun [36] and Wagenaar and Groeneweg [40], a Dutch study of 100 marine casualties reported that the occurrence of human error was a factor in 96 of the 100 accidents.

Additional information on shipping system human error-related facts and figures is available in the study by Dhillon [27].

6.11 Marine Industry-Related Human Factors Issues and Methods for Reducing the Manning Impact on Shipping System Reliability

There are many human factors issues that can directly or indirectly influence the occurrence of human errors in the marine industry. Some of these issues are as follows [36,41,42]:

- Poor communication
- Poor maintenance
- Faulty policies, practices, or standards
- Fatigue
- Poor knowledge of the ships' systems
- Hazardous natural environment
- Poor general technical knowledge
- Poor automation design
- Decisions based on inadequate information

Additional information on the nine issues previously mentioned is available in the study by Dhillon [27].

Three methods considered useful for reducing the manning impact on shipping system reliability in regard to improving human reliability are as follows [43]:

Method 1: Reduce the occurrence of human error incidence—In this case, the occurrence of human errors is reduced through actions such as the following:

Simplification of job task

Application of human engineering design principles

Human error occurrence likelihood analysis or modeling

Method 2: Improve mean time between failures under the reduced manning environment—In this case, the mean time between failures is improved through actions such as the following:

Choosing/designing highly reliable system parts/components

Designing the interfaces to optimize the use of these parts/components

Method 3: Minimize or eliminate human error impacts—In this case, human error impacts are minimized or eliminated through actions such as the following:

Designing the system to be fully error tolerant

Designing the system that clearly enables the system/human to recognize that an error has occurred and to correct the error before the occurrence of any damage

Additional information on these methods is available in the study by Anderson et al. [43].

6.12 Road Transportation System Human Error-Related Facts and Figures and Common Driver Errors

Some of the facts and the figures directly or indirectly concerned with human error in road transportation systems are as follows:

- As per the study by Jacobs et al. [2], in five developing countries (i.e., India, Tanzania, Thailand, Nepal, and Zimbabwe), during the period of 1966–1998, over 70% of bus-related accidents were the result of driver error.
- As per the study by South African Press Association [44], approximately 65% of motor vehicle-related accidents are due to human error.
- As per the study by *The Detroit News* [45], approximately 57% of bus accidents in South Africa are due to human error.
- As per www.driveandsurvive.ca.uk/cont5.htm [46], human error is cited more often than mechanical-related problems in approximately 5000 truck-related fatalities that occur annually in the United States.
- A study concerning transportation safety issues reported that most of the truck–car crashes were due to human error committed by either the truck driver or the car driver [47].

- As per the studies by Krug [48], Odero [49], and Murray and Lopez [50], the annual cost of worldwide road crashes is around $500 billion, and by the year 2020, it is predicted that road traffic-related injuries will become the third largest cause of disabilities in the world.

During the driving process, drivers make various types of errors. Some of the common ones are as follows [51,52]:

- Following too closely
- Overtaking or passing in the face of incoming traffic
- Changing lanes abruptly
- Following closely prior to overtaking
- Overtaking at a junction or a crossroad
- Straddling lanes
- Following closely a motor vehicle that is overtaking
- Driving too fast for prevailing circumstances

6.13 Classifications and Ranking of Driver Errors

Motor vehicle drivers make various types of errors. These errors may be categorized under four classifications. These classifications, with respect to the decreasing frequency of occurrence, are as follows [53,54]:

- *Classification 1:* Recognition errors
- *Classification 2:* Decision errors
- *Classification 3:* Performance errors
- *Classification 4:* Miscellaneous errors

Additional information on the previously mentioned classifications of errors is available in the studies by Rumar [53] and Treat [54].

Over the years, many studies have been carried out to rank the occurrence of the driver errors. The findings of one of these studies that ranked driver errors/causes from the lowest frequency of occurrence to the highest frequency of occurrence are presented in the following [52]:

- Poor skill
- Faulty signaling

- Lack of education or road craft
- Incorrect decision/action
- Reckless or irresponsible
- Difficult maneuver
- Following too closely
- Misjudged distance and speed
- Lack of judgment
- Wrong interpretation
- Improper overtaking
- Poor attention
- Wrong path
- Failure to look
- Inexperience
- Distraction
- Looked but failed to see
- Too fast
- Lack of care

PROBLEMS

1. Discuss the following two items:
   a. Brake defects
   b. Rim defects
2. Describe rail weld failures.
3. What are the classifications of road and rail tanker failure modes? Discuss each of these classifications in detail.
4. Discuss the mechanical failure-related aviation accidents and give at least four examples of such accidents.
5. Discuss ship failures and common causes for their occurrence.
6. What are the typical human error occurrence areas in railway operation? Describe each of these areas in detail.
7. List at least five aviation system human error-related facts and figures.
8. What are the useful recommendations to reduce communication-related errors between pilots and controllers?
9. What are the common driver errors?
10. List at least five road transportation system human error-related facts and figures.

References

1. International Air Transport Association, IATA 2006 fast facts: The air transport industry in Europe has united to present its key facts and figures, International Air Transport Association, Montreal, QC, retrieved on March 10, 2008. Available online at www.iata.org/pressroom/economics.facts/stats/2003-04-10-01.htm.
2. Jacobs, G., Aeron-Thomas, A., Astrop, A., *Estimating Global Road Fatalities*, Report No. TRL 445, Transport Research Laboratory, Wokingham, 2000.
3. Pearce, T., Maunder, D.A.C., *The Causes of Bus Accidents in Five Emerging Nations*, Report, Transport Research Laboratory, Wokingham, 2000.
4. Anderson, J.E., *Transit Systems Theory*, D.C. Heath, Lexington, MA, 1978.
5. Dhillon, B.S., *Transportation Systems Reliability and Safety*, CRC Press, Boca Raton, FL, 2011.
6. Fukada, Y., Yamamoto, R., Harasawa, H., Nakanowatari, H., Experience in maintaining rail track in Japan, *Welding in the World*, Vol. 47, 2003, pp. 123–137.
7. Tatsumi, M., Fukada, Y., Veyama, K., Shitara, H., Yamamoto, R., Quality evaluation methods for rail welds in Japan. *Proceedings of the World Congress on Railway Research*, 1997, pp. 197–205.
8. Cannon, D.F., Edel, K.O., Grassie, S.L., Sawley, K., Rail defects: An overview, *Fatigue and Fracture of Engineering Materials and Structures*, Vol. 26, No. 10, 2003, pp. 865–886.
9. Labropoulos, K.C., Moundoulas, P., Moropoulou, A., Methodology for monitoring, control and warning of defects for preventive maintenance of rails, in *Computers in Railways X*, WIT Transactions on the Built Environment, WIT Press, London, 2006, pp. 935–944.
10. Marshall, V.C., Modes and consequences of the failure of road and rail tankers carrying liquefied gases and other hazardous liquids, in *Reliability on the Move*, edited by G.B. Guy, Elsevier Science, London, 1989, pp. 136–148.
11. *Aircraft Accident Report: United Airlines Flight 585*, Report No. AAR92-06, National Transportation Safety Board, WA, DC, 1992. Available online at www .libraryonline.erau.edu/online-full-text/ntsb/aircraft-accident-reports /AAR92-06.pdf.
12. Byrne, G., *Flight 427: Anatomy of an Air Disaster*, Springer-Verlag, New York, 2002.
13. *Report on the Accident to Boeing Vetrol (BV) 234LR, G-BWFC 2.5 Miles East of Sumburgh, Shetland Isles*, November 6, 1986, Report No. 2, Air Accidents Investigation Branch, Berkshire Copse Road, Aldershot, 1988.
14. Johnston, M., *The Last Nine Minutes: The Story of Flight 981*, Morrow Publisher, New York, 1976.
15. Gero, D., *Aviation Disasters*, Patrick Stephens Limited, Sparkford, 1993.
16. United Airlines Flight 859, Aircraft Accident Report No. SA-362 (file 1-0003), Civil Aeronautics Board, Washington, DC, 1962.
17. Reinach, S., Viale, A., Application of a human error framework to conduct train accident/incident investigations, *Accident Analysis and Prevention*, Vol. 38, 2006, pp. 396–406.
18. Anderson, T., Human reliability and railway safety. *Proceedings of the 16th European Safety, Reliability, and Data Association (ESREDA) Seminar on Safety and Reliability in Transportation*, 1999, pp. 1–12.

19. *The Nation Newspaper*, Human error derails New Metro, Editorial, Bangkok, January 18, 2005.
20. Hall, S., *Railway Accidents*, Ian Allan Publishing, Shepperton, 1997.
21. Railway Board of the Ministry of Railways, *Safety on Indian Railways* [White paper], Railway Board, Ministry of Railways, Government of India, New Delhi, April, 2003.
22. Hudoklin, A., Rozman, V., Reliability of railway traffic personnel, *Reliability Engineering and System Safety*, Vol. 52, 1996, pp. 165–169.
23. Dhillon, B.S., *Human Reliability and Error in Transportation Systems*, Springer-Verlag, London, 2007.
24. Report No. 1–96, *Statistical Summary of Commercial Jet Accidents: Worldwide Operations: 1959–1996*, Boeing Commercial Airplane Group, Seattle, WA, 1996.
25. Mjos, K., Communication and operational failures in the cockpit, *Human Factors and Aerospace Safety*, Vol. 1, No. 4, 2001, pp. 323–340.
26. *Science Daily*, Fewer airline crashes linked to "pilot error," inclement weather still major factor, January 9, 2001.
27. Dhillon, B.S., *Safety and Human Error in Engineering Systems*, CRC Press, Boca Raton, FL, 2013.
28. Helmreich, R.L., *Managing Human Error in Aviation*, Scientific American, 1997, May, pp. 62–67.
29. Cardosi, K., Falzarano, P., Han, S., *Pilot–Controller Communication Errors: An Analysis of Aviation Safety Reporting System (ASRS) Reports*, Report No. DOT/FAA/AR-98/17, Federal Aviation Administration (FAA), Washington, DC, August, 1998.
30. Von Thaden, T.L., Wiegmann, D.A., Shappell, S.A., Organization factors in commercial aviation accidents 1990–2000. Paper presented at the 13th International Symposium on Aviation Psychology, Dayton, OH, 2005.
31. Just waiting to happen—The work of the UK P$I Club, *The International Maritime Human Element Bulletin*, No. 1, October 2003, pp. 3–4.
32. Boniface, D.A., Bea, R.G., Assessing the risks of and countermeasures for human and organizational error, *SNAME Transactions*, Vol. 104, 1996, pp. 157–177.
33. *Asia Maritime Digest*, DVD spotlights human error in shipping accidents, January/February 2004, pp. 41–42.
34. *The Scandinavian Shipping Gazette*, Human Error Led to the Sinking of the Bow Mariner, 2006.
35. *Working Paper on Tankers Involved in Shipping Accidents 1975–1992*, Transportation Safety Board of Canada, Ottawa, ON, 1993.
36. Rothblun, A.M., Human error and marine safety. *Proceedings of the Maritime Human Factors Conference*, Linthicum, MD, 2000, pp. 1–20.
37. Bryant, D.T., *The Human Element in Shipping Causalities*. Report prepared for the Department of Transport, Marine Directorate, London, 1991.
38. Hee, D.D., Pickrell, B.D., Bea, R.G., Roberts, K.H., Willliamson, R.B., Safety Management Assessment System (SMAS): A process for identifying and evaluating human and organization factors in marine system operations and filed test results, *Reliability Engineering and System Safety*, Vol. 65, 1999, pp. 125–140.
39. Moore, W.H., Bea, R.G., *Management of Human Error in Operations of Marine Systems*, Report No. HOE-93-1, 1993. Available from the Department of Naval Architecture and Offshore Engineering, University of California, Berkeley, CA.

40. Wagenaar, W.A., Groeneweg, J., Accidents at sea: Multiple causes and impossible consequences, *International Journal of Man-Machine Studies*, Vol. 27, 1987, pp. 587–598.
41. McCallum, M.C., Raby, M., Roghblum, A.M., *Procedures for Investigating and Reporting Human Factors and Fatigue Contributions to Marine Casualties*, US Coast Guard Report No. CG-D-09-07, Department of Transportation, Washington, DC, 1996.
42. *Human Error in Merchant Marine Safety*, Report by the Marine Transportation Research Board, National Academy of Science, Washington, DC, 1976.
43. Anderson, D.E., Malone, T.B., Baker, C.C., Recapitalizing the navy through optimized manning and improved reliability, *Naval Engineers Journal*, November 1998, pp. 61–72.
44. *Driving Related Facts and Figures, UK*, July 2006. Available online at www.drive andsurvive.ca.uk/cont5.htm.
45. *Poor Bus Accident Record for Gauteng*, South African Press Association (SAPA), Cape Town, 2003, July 4.
46. *The Detroit News*, Trucking safety snag: Handling human error, 2000, July 17.
47. Zogby, J.J., Knipling, R.R., Werner, T.C., *Transportation Safety Issues*, Report by the Committee on Safety Data, Analysis and Evaluation, Transportation Research Board, Washington, DC, 2000.
48. Krug, E., ed., *Injury: A Leading Cause of the Global Burden of Disease*, World Health Organization, Geneva, 1999.
49. Odero, W., Road traffic injury research in Africa: Context and priorities. *Paper at the Global Forum for Health Research Conference (Forum 8)*, November, 2004.
50. Murray, C.J.L., Lopez, A.D., *The Global Burden of Disease*, Harvard University Press, Boston, MA, 1996.
51. Harvey, C.F., Jenkins, D., Sumner, R., *Driver Error*, Report No. TRRL SR 149, Transport and Research Laboratory, Department of Transport, Crowthorne, 1975.
52. Brown, I.D., Drivers' margin of safety considered as a focus for research on error, *Ergonomics*, Vol. 33, 1990, pp. 1307–1314.
53. Rumar, K., The basic driver error: Late detection, *Ergonomics*, Vol. 33, 1990, pp. 1281–1290.
54. Treat, J.R., *A Study of Pre-crash Factors Involved in Traffic Accidents*, Report No. HSRI 10/11,6/1, Highway Safety Research Institute (HSRI), University of Michigan, Ann Arbor, MI, 1980.

7

Software, Robot, and Transportation System Safety

7.1 Introduction

Today, computers have become an important element of day-to-day life, and each year, billions of dollars are spent to develop various types of software, and the safety of software has become an important issue. More specifically, in many applications, the proper functioning of software is so critical that a simple malfunctioning can result in a large-scale loss of lives and in a high cost. For example, commuter trains in Paris, France, each day, serve around 800,000 passengers and depend on software signaling [1].

Nowadays, robots are being used in many diverse areas and applications, and their safety-related problems have significantly increased over the years. The history of robot safety may be traced back to the early years of the 1980s with the development of the Japanese Industrial Safety and Health Association document entitled "An Interpretation of the Technical Guidance on Safety Standards in the Use, etc., of Industrial Robots" and the American National Standard for Industrial Robots and Robot Systems: Safety Requirements [2,3].

Over the years, the safety of transportation systems has become a very important issue. For example, in 1990, there were about 1 million traffic-related deaths and approximately 40 million traffic-related injuries around the globe, and as per the projection of the World Health Organization, global deaths from accidents will increase to about 2.3 million by 2020 [4,5].

This chapter presents various important aspects of software, robot, and transportation system safety.

7.2 Software Potential Hazards and Software Risk and Safety Classifications

There are a number of ways in which software can cause/contribute to a hazard. Some of these are as follows [6–8]:

- Failed to carry out a required function
- Performed a function out of sequence
- Responded poorly to a contingency
- Provided incorrect solution to a problem
- Carried out a function not required
- Failed to recognize a hazardous condition requiring a corrective action

There are basically three ways in which software can increase risk. These are directing the system toward a hazardous direction or state, failure to mitigate the damage after an accident, and failure to detect and take an appropriate corrective action to recover from a hazardous situation [9]. Software risks, in which losses can occur, may be categorized under three classifications as follows [10,11]:

- *Classification A:* This classification includes situations such as poorly funded or planned projects, unavailability of required manuals, and poorly trained workers to perform the assigned tasks.
- *Classification B:* This classification includes situations where environmental conditions may, directly or indirectly, impact a software professional's ability to perform his/her job effectively. Some examples of such conditions are screen glare, incorrect software development tools for the job, poor lighting, and inadequate computer memory/hardware.
- *Classification C:* This classification includes situations such as no usage of standards, inadequate standards, and no compliance to procedures and policies.

Software safety may be grouped under three classifications as follows [12]:

- *Safety software:* This type of software controls or performs such functions, which are activated to prevent or minimize the effect of a safety-critical system failure.
- *Nonsafety software:* This type of software controls or performs system functions that are not concerned with safety.

- *Safety-critical software:* This type of software controls or performs such functions; if they are executed erroneously or if they failed to execute correctly, they could directly inflict serious injuries to humans and/or the environment and cause deaths.

Finally, it is to be noted that most mission-critical systems incorporate a combination of safety, nonsafety, and safety-critical software.

7.3 Software System Safety-Associated Tasks and Role of Software Quality Assurance Organization with Respect to Software Safety

There are many software system safety-associated tasks. Nine of these tasks are presented in the following [8,13]:

- Develop safety-associated software test case requirements, test plans, test descriptions, and test procedures.
- Identify all safety-critical elements and variables for use by code developers.
- Review all the test results concerning safety issues and trace all the highlighted safety-associated software shortcomings back to the system level.
- Highlight software elements that, directly or indirectly, control safety-critical operations and then direct safety analysis and tests on those specific functions as well as on the safety-critical path that leads to their execution.
- Trace highlighted system-related hazards to the hardware–software interface.
- Develop an appropriate tracking system within the software along with system configuration control structure for ensuring the traceability of safety-related requirements and their flow through documentation.
- Show the software system safety constraint-associated consistency in regard to the software requirement specification.
- Develop appropriate system-specific software design-related requirements and criteria, computer–human interface-associated requirements, and testing requirements on the basis of highlighted software system safety-associated constraints.
- Trace safety-related constraints and requirements right up to the code level.

A software quality assurance organization plays various roles with respect to software safety. Some of these roles are presented in the following [8,14]:

- Conduct safety audits and reviews of operational systems regularly.
- Define the user safety-associated requirements, the operational concept, and the operational doctrine.
- Define the appropriate requirements for performing operational safety-related reviews.
- Approve the findings of safety testing before releasing the systems.
- Develop the operational safety-related policy that identifies acceptable risks as well as operational alternatives to hazardous operations.
- Investigate, evaluate, resolve, and document the reported safety-related operational incidents.
- Determine the necessary safety-related criteria for system acceptance.

7.4 Software Safety Assurance Program

A software safety assurance program within an organizational setup basically involves three maturity levels presented in the following [8,11,14].

- *Maturity level 1:* This maturity level is concerned with the development of the culture of an organization/company that clearly recognizes the importance of issues concerning safety. More clearly, all company software developers carry out their tasks as per the standard development rules and apply them consistently.
- *Maturity level 2:* This maturity level is concerned with the implementation of an appropriate development process that involves safety assurance reviews as well as hazard analysis to identify and eliminate safety-critical situations before being designed into the system.
- *Maturity level 3:* This maturity level is concerned with the utilization of an appropriate design process that documents results as well as implements continuous improvement methods for eliminating safety-critical errors in the system software.

It is to be noted that some of the items that need to be carefully considered during the implementation process of a software safety program are as follows [8,11]:

- Software system safety-related requirements are specified and developed as an element of the design policy of the organization.

- Software system policy is clearly quantifiable to the stated risk level using the general measuring methods.
- All human–computer interface requirements as well as software system safety requirements are clearly consistent with contract requirement.
- All changes in design, mission requirements, or configuration are performed such that they clearly maintain an acceptable level of risk.
- Software system safety is addressed in regard to team effort that clearly involves groups such as management, engineering, and quality assurance.
- Past software safety-related data are carefully considered and used in all future software development projects.
- All software system-associated hazards are tracked, identified, evaluated, and eliminated as per requirements.

7.5 Software Hazard Analysis Methods

There are many methods that can be used to perform various types of software hazard analysis. Ten of these methods are presented in Table 7.1 [8,11,15–18].

The first three of the methods presented in Table 7.1 are described in the following, and the additional information on the remaining seven methods is available in the studies by Dhillon [8,11], Ippolito and Wallace [15], Sheriff [16], Hansen [17], and Hammer and Price [18].

TABLE 7.1

Methods for Performing Software Hazard Analysis

No.	Method Name
1	Software sneak circuit analysis
2	Code walk-through
3	Proof of correctness
4	Software FTA
5	Software/hardware integrated critical path
6	Petri net analysis
7	Cause–consequence diagrams
8	Hazard and operability studies
9	FMEA
10	Event tree analysis

7.5.1 Software Sneak Circuit Analysis

Software sneak circuit analysis is a method that is concerned with identifying software logic that leads to the occurrence of undesirable outputs. The program source code is converted to topological network trees, and six basic patterns are used for modeling the code: entry dome, return dome, iteration loop, trap, single line, and parallel line. Each software mode is modeled with the aid of these basic patterns linked in a network tree flowing from top to bottom.

The involved person (i.e., analyst) asks questions with respect to interrelationships and use of the instructions that are considered as the structural elements/components. The clear answers to the questions asked provide clues that highlight sneak conditions that may result in undesirable outputs.

The involved analysts search for the following basic software sneaks:

- Existence of an undesired output
- A program message poorly describing the very actual condition
- Wrong timing
- The undesired inhibit of an output

The clue-generating questions are taken from the topograph denoting the code segment, and whenever sneaks are found, the involved analysts conduct investigative analyses to verify that the code indeed produced the sneaks. Subsequently, the sneaks' impacts are assessed and appropriate corrective measures are recommended. Additional information on this method is available in the study by Ericson [19].

7.5.2 Code Walk-Through

Code walk-through is a method that is considered quite useful for improving software safety and is basically a team effort among professionals such as software engineers, software programmers, system safety persons, and program managers. Code walk-throughs entail a quite rigorous review of the software through inspection and discussion of the software functionality. All logic branches as well as the function of each and every statement are discussed quite thoroughly.

The system reviews the software functionality and compares it with the system-related requirements. This verifies that all software-associated requirements are implemented properly, in addition to determining the functionality accuracy. Additional information on this method is available in the study by Sheriff [16].

7.5.3 Proof of Correctness

Proof of correctness is quite a useful method for performing software hazard analysis, and it decomposes a program into a number of logical segments.

In turn, for each of these segments, input/output assertions are defined. Subsequently, the involved software professional verifies with care from the perspective that each and every input assertion and its output assertion are true and that if all input assertions are true, then all output assertions are also true.

Finally, it is to be noted that the proof-of-correctness method uses mathematical theorem-proving concepts to verify that the program in question is consistent with its specifications. Additional information on this method is available in the study by Weik [20].

7.6 Robot Hazards and Safety-Related Problems

The following are the three basic types of robot hazards [21–24]:

- *Impact hazards:* This type of hazard involves being struck by the moving parts of a robot or by the item/parts being carried by the robot. These hazards include being struck by flying objects that are dropped or ejected by the robot.
- *Trapping hazards:* This type of hazard is generally the result of robot movements with respect to fixed objects such as machines and pests in the same area. The movements of the auxiliary equipment could be the other possible causes. Two examples of such equipment are carriages and pallets.
- *Hazards that develop from the application:* Some examples of the hazards that develop from the application are electric shocks, exposure to toxic substances, arc flash, and burns. The most prevalent causes of hazards such as these are control errors, mechanical-related problems, human errors, and unauthorized access [25].

Over the years, it has been observed that safety professionals concerned with robots face many unique robot safety problems. Some of these problems are as follows [21–24]:

- Generally, robots operate closely with other machines and humans in the same environment. In particular, humans are subject to colliding with the moving parts of robots, tripping over loose control/power cables (if any), and being pinned down.
- Robots generate potentially hazardous conditions/situations because they manipulate objects of varying sizes and weights.
- A robot may lead to quite a high risk of fire or explosion if it is placed in a rather unfriendly environment.

- The presence of a robot receives a rather high attention from people in the surrounding area, who are often ignorant of the possible associated hazards.
- In the event of a hydraulic, a control, or a mechanical failure, robots may move out of their programmed area zones and strike something or they may throw some small object.
- Robot mechanical design-associated problems may lead to hazards such as pinching, grabbing, and pinning.
- Robot-related maintenance procedures may result in hazardous situations.
- Various safety-related electrical design problems can occur in robots/ robot systems. Some examples of these problems are potential for electric shock, poorly designed power sources, and fire hazards.

7.7 Robot Safety-Related Problems Causing Weak Points in Planning, Design, and Operation

There are many weak points in planning, design, and operation, which can lead to robot-related safety problems in an industrial setting [3,24,25]. Some of the weak points concerned with planning are shown in Figure 7.1. The weak points shown in the figure are poor spatial arrangement, poor organization

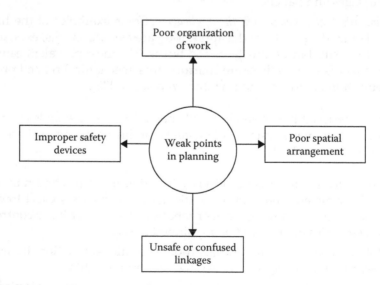

FIGURE 7.1
Weak points in planning with respect to robot safety.

of work, unsafe or confused linkages, and improper safety devices. Poor spatial arrangement can lead to confusion and the possibility of collision. Poor organization of work is an important factor, particularly in programming and stoppages. Unsafe or confused linkages are basically concerned with interfaces between individual machines. Finally, improper safety devices are composed of improper guards (i.e., being too low, containing gaps, or being close to hazard points) and faulty emergency shutdown circuits.

Some of the weak points concerned with design are as follows [26]:

- Poor gripper design, specifically when the power fails
- Poor cable and hose strength, as well as poor laying
- Poor defense against unintentional activation of operating devices
- Trivial control errors which lead to hazardous system states
- Poor design in regard to human factors
- Poor defense against environment influences (e.g., temperature, dust, and swarf)
- Incapacitation of primary safety devices
- Part failure which leads to hazardous system states

Finally, some of the weak points in operational procedure in regard to robot safety in industrial settings are presented in the following:

- Failure to provide proper feedback to individuals involved in design and layout concerning weak spots and how to ensure their removal
- Allowing countersafety working procedures during a stoppage
- Poor training of workers who are involved with industrial robots

---

7.8 Common Robot Safety-Related Features and Their Functions

Generally, a well-designed robot incorporates, to the greatest extent possible, safety-related features that take into consideration all modes of robot operation (i.e., normal operation, maintenance, and programming) [27,28]. Nonetheless, some of the safety features are usually common to all robots, and the others are specific to the types of robots. Some of the common robot safety-related features, along with their corresponding intended functions, given in parentheses, are presented in the following [27]:

- Stop button (it removes manipulator and control power)
- Hold/run button (it stops arm motion but leaves power on)

- Slow-speed control (it allows program execution at reduced speeds)
- Hardware stops (complete control on travel/movement limits)
- Condition indicators and messages (these provide visual indication by lights/display screens of the system condition)
- Hydraulic fuse (it protects against high-speed motion/force in teach mode)
- Power disconnect switch (it removes total power at the machine junction box)
- Servomotor brake (it maintains the position of the arm at a standstill)
- Software stops (computer-controlled travel/limit)
- Remote connections (these allow remote control of necessary machine/safety functions)
- Teach pendant trigger (it must be held by the involved operator for arm power in teach mode)
- Error detection, parity checks, etc. (whereby the computer approaches for self-checking of a variety of functions)
- Teach/playback mode selector (it provides the involved operator with control over the operating mode)
- Line indicator (it indicates that incoming power is connected at the junction box)
- Arm-power-only button (it applies power only to the manipulator)
- Step button (it allows program execution one step at a time)
- Automatic/manual dump (it provides means for relieving hydraulic/pneumatic pressure)
- Control-power-only button (it applies power only to the control section)

7.9 Robot Safeguard Methods

There are many robot safeguard methods [23]. Three of these methods are described in the following, separately [23–25].

7.9.1 Flashing Lights

The flashing light method calls for the installation of a flashing light on the robot itself or at the perimeter of the work area of the robot. The purpose of the flashing light is to alert humans in the area that programmed motion is in progress or could happen any time.

It is to be noted that when the flashing light method is used, ensure that the flashing light is energized continuously during the period when the robot drive power is activated.

7.9.2 Intelligent Systems

These systems make their decisions through remote sensing, software, and hardware. In order to achieve an effective intelligent collision avoidance system, the operating environment of a robot has to be restricted, in addition to the wide use of special sensors and software. This calls for the need for a sophisticated computer to make correct decisions and real-time computations.

Finally, it is to be noted that in most industrial settings, usually, it is not possible to restrict the environment effectively.

7.9.3 Warning Signs

Warning signs are used in situations where robots, by virtue of their speed, size, and inability to impart significant force, cannot injure people. Two examples of such situations are small-part assembly and laboratory robots. Robots such as these need no special safeguarding as warning signs are sufficient for the uninformed individuals in the area.

However, it is to be noted that warning signs are useful for all robot application areas, irrespective of whether robots possess the ability of injuring people or not.

7.10 Truck Safety-Related Facts and Figures

Some of the truck safety-related facts and figures are as follows:

- As per the study by Cox [29], in 1993, about 80% of truck accidents in the United States occurred with no adverse weather conditions.
- As per the study by Cox [29], around 65% of large truck crash fatalities in the United States occur on major roads.
- In 2000, 1997, 1995, 1992, 1989, 1986, and 1980, there were approximately 5000, 4900, 4500, 4000, 5000, 5100, and 5400 truck-related fatal crashes in the United States, respectively [30].
- As per the study by Sheiff [31], during the period from 1976 to 1987, the fatalities of truck occupants decreased from 1130 in 1976 to 852 in 1987 in the United States.

- As per the studies by the Transportation Research Board [30] and the Federal Motor Carrier Safety Administration [32], in 2003, in the United States, out of a total of 4986 deaths that occurred from crashes involving large trucks, 78% were occupants of another vehicle, 14% were occupants of large trucks, and 8% were not occupants.
- As per the study by Cox [29], in 1993, in the United States, approximately 4500 trucks were involved in accidents in which at least 1 fatality occurred.

7.11 Truck and Bus Safety-Related Issues

Over the years, many studies have been performed to highlight truck and bus safety-related issues. Some of the important ones are as follows [33,34]:

- *Fatigue:* This issue is concerned with scheduling, driving, unloading, and road conditions that induce it, in addition to hours-of-service violations and lack of proper places for rest.
- *Working conditions:* This issue is concerned with the review of ongoing industry practices and standards as they affect drivers' workload.
- *Enforcement:* This issue is concerned with the need for better traffic-related enforcement, licensing and testing, and adjudication of highway user violations.
- *Driver training:* This issue is concerned with the need for continuing and better education for all involved drivers (i.e., commercial and private).
- *License deficiencies:* This issue is concerned with the review of testing procedures followed for licenses of commercial drivers.
- *Resource allocations:* This issue is concerned with the priorities and the allocation of all scarce resources through a better and effective safety management system that clearly gives safety top priority.
- *Uniform regulations:* This issue is concerned with the lack of uniformity in safety-related regulations and procedures among the states and between Canada and Mexico, clearly indicating that safety-related issues do not receive the same degree of priority in all involved jurisdictions.
- *Accident countermeasures:* This issue is concerned with the proper research efforts targeted for seeking and defining proactive and nonpunitive countermeasures for preventing accidents.

- *Communications:* This issue is concerned with the development of an effective national motor carrier safety-related campaign as well as the expansion of education-related efforts to the public at large for properly sharing roads with large vehicles.
- *Technology:* This issue is concerned with the development and the deployment of proper emerging and practically inclined technologies for improving safety.
- *Partnership:* This issue is concerned with better and effective communication and coordination among all highway users.
- *Information/data:* This issue is concerned with the shortage of information on heavy vehicle-related crashes and their causes.

7.12 Recommendations for Improving Truck Safety

In 1995, the attendees of a conference on "Truck Safety: Perceptions and Reality" made many useful recommendations on the following five issues for improving truck safety [35,36]:

- Driver training and empowerment
- Driver fatigue
- Vehicle brakes and maintenance standards
- Harmonization of safety standards across all jurisdictions
- Data needs

Recommendations on each of the five issues previously mentioned are presented in the following sections, separately.

7.12.1 Recommendations on Driver Training and Empowerment Issue

The following four recommendations were made on the driver training and empowerment issue [35]:

1. Enact an accreditation of all involved driver training schools for ensuring that they uniformly satisfy required standards in all jurisdictions.
2. Devise and enforce necessary regulations for ensuring that truck drivers are not unfairly dismissed when they refuse to drive in unsafe conditions.
3. Implement an appropriate graduated licensing scheme that clearly reflects the need of a variety of trucking vehicles.

4. Establish appropriate driver training and retraining programs that clearly focus on safety (e.g., teaching drivers on how to effectively inspect the vehicle by utilizing the latest technology) and take appropriate measures for reducing accident risk.

7.12.2 Recommendations on Driver Fatigue Issue

The following three recommendations were made on the driver fatigue issue [35]:

1. Set appropriate tolerance levels for accident risk and fatigue and devise new standards that clearly incorporate these levels.
2. Develop a comprehensive approach for highlighting the incidence of fatigue of truck drivers that effectively takes into consideration different types of fatigue and driving-related requirements.
3. Harmonize all applicable standards across different jurisdictions.

7.12.3 Recommendations on Vehicle Brakes and Maintenance Standards Issue

The following four recommendations were made on the vehicle brakes and maintenance standards issue [35]:

1. Train and certify all involved truck drivers to adjust vehicle brakes properly as part of their licensing requirements and training program.
2. Equip trucks with essential onboard devices/signals for indicating when brakes require adjustment and servicing.
3. Invoke appropriate penalties for those companies/organizations that regularly fail to satisfy required inspection standards.
4. Implement an appropriate safety rating system.

7.12.4 Recommendations on Harmonization of Safety Standards across All Jurisdictions Issue

The following two recommendations were made on the harmonization of safety standards across all jurisdictions issue [35]:

1. Establish an appropriate agency for collecting and disseminating safety-related information to all concerned parties.
2. Form a committee of government and industry representatives to explore avenues for cooperative efforts for developing uniform truck safety-related standards.

7.12.5 Recommendations on Data Needs Issue

The following four recommendations were made on the data needs issue [35]:

1. Highlight and share currently available truck accident and exposure-related data.
2. Standardize all accident-reporting forms being used by police in all jurisdictions.
3. Establish a North American truck safety-related data center.
4. Improve the reliability of police accident-related reports through better police training for collecting and reporting reliable data concerning accident causes and consequences.

7.13 Examples of Rail Accidents and Their Causes

Over the years, there have been many rail-related accidents around the globe due to various causes. Six examples of such accidents are as follows:

- *Example 1:* In 2004, in Macdona, Texas, United States, a Union Pacific Railway train failed to stop at a signal and collided with another train that resulted in 3 fatalities and 51 injuries [37].
- *Example 2:* In 2002, near Crescent City, Florida, United States, an Amtrak autotrain derailed because of malfunctioning brakes and poor track maintenance and resulted in 4 fatalities and 142 injuries [38].
- *Example 3:* In 1999, in Waipahi, New Zealand, a northbound Main South Line express freight train collided with a stationary southbound freight train because of misunderstanding of track warrant conditions by both train drivers and resulted in 1 fatality and 1 serious injury [39].
- *Example 4:* In 1957, in Dundrum, Ireland, a passenger train delayed by a cow on the line was struck from behind by another passenger train mistakenly signaled into the station and resulted in 1 fatality and 4 serious injuries [40,41].
- *Example 5:* In 1943, in Hyde, New Zealand, a Cromwell-to-Dunedin passenger train derailed on a curve because of excessive speed due to an intoxicated driver and resulted in 21 fatalities and 47 injuries [39].
- *Example 6:* In 1864, in Ballinasloe, Ireland, a passenger train derailed because of excess speed on a poor track and resulted in 2 fatalities and 34 injuries [42].

7.14 Classifications of Rail Accidents by Causes and Effects

Rail accidents by causes can be grouped under six classifications. These classifications are as follows [43–45]:

- *Classification 1: Signalperson errors*—This classification includes causes such as allowing two trains into the same occupied block section and incorrectly operating signals, points, or token equipment.
- *Classification 2: Mechanical failure of rolling stock*—This classification includes causes such as poor design and maintenance.
- *Classification 3: Driver errors*—This classification includes causes such as excessive speed, engine mishandling, and passing signals at danger.
- *Classification 4: Civil engineering failure*—This classification includes causes such as bridge and tunnel collapses and track (permanent way) faults.
- *Classification 5: Contributory factors*—This classification includes causes such as poor track or junction layout, effectiveness of brakes, rolling stock strength, and poor rules.
- *Classification 6: Acts of other people*—This classification includes causes such as the acts of other railway personnel and of nonrailway personnel (i.e., vandalism, accidental damage, and terrorism). Two examples of other railway personnel are porters and shunters (workers who couple and uncouple cars).

Three commonly proposed classifications of rail accidents by effects are shown in Figure 7.2 [43–45].

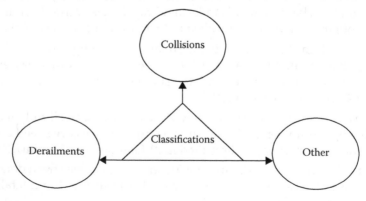

FIGURE 7.2
Classifications of rail accidents by effects.

The *derailment* classification includes items such as plain track, junctions, and curves. Similarly, the *collision* classification includes items such as head-on collisions, collisions with buffer stops, near collisions, and obstructions on the track/line (i.e., avalanches, road vehicles, landslides, etc.). Finally, the *other* classification includes items such as collisions with people on tracks, falls from trains, and fire and explosions.

7.15 Railroad Tank Car Safety

Railroad tank cars are used for transporting various types of liquids and gases from one point to another, and their contents are flammable and corrosive or pose other various hazards if released accidently. In the United States alone, during the period of 1965–1980, tank car accidents resulted in 40 fatalities, and accidental releases take place roughly once out of every 1000 shipments and result in approximately 1000 releases annually [46].

In order to ensure the safety of tank cars, in the 1990 Hazardous Materials Transportation Uniform Safety Act, the US Congress called for the review of the tank car design process and an assessment of whether head shields should be made mandatory on all types of railroad tank cars used to transport hazardous materials [46]. In turn, to address both these issues, the Transportation Research Board formed a committee of experts in areas such as tank car design, transportation and hazardous material safety, railroad operations and labor, transportation regulations and economics, and chemical shipping.

After examining railroad tank car incident-related data, the committee made three recommendations presented in the following [46].

- *Recommendation 1:* Improve the information and the criteria employed to assess the safety-related performance of tank car design types as well as to assign materials to tank cars.

- *Recommendation 2:* Improve the cooperation between the industry and the Department of Transportation to identify critical safety-associated needs and take necessary action for achieving them.

- *Recommendation 3:* Improve the implementation process of industry design approval and certification function as well as all federal oversight procedures and processes.

7.16 Analysis of World Airline Accidents

Nowadays, airlines have become a widely used mode of transportation throughout the world, and over 16,000 jet aircraft are being used [47]. A study of worldwide scheduled commercial jet operations during the period of 1959–2001 revealed that there were 1307 accidents that resulted in 24,700 onboard deaths [47,48]. The three classifications of the 1307 accidents along with their number and percentage (in parentheses) are as follows [47,48]:

- Passenger operations: 1033 accidents (79%)
- Cargo operations: 169 accidents (13%)
- Training, demonstration, testing, or ferrying: 105 accidents (8%)

The collective Canadian and US proportion of these 1307 accidents was around 34% (i.e., 445 accidents), resulting in about 25% (i.e., 6077) of the 24,700 onboard deaths [47].

A study of the accident data for the previously mentioned period (i.e., 1959–2001) also revealed that the accident rate (i.e., accidents per million departures) for the period was fairly stable [48]. Additional information on the topic is available in the studies by Wells and Rodrigues [47] and the Boeing Commercial Airplane Company [48].

7.17 US Airline-Related Fatalities and Causes of Airplane Crashes

The history of airline-related fatal accidents in the United States goes back to the later years of the 1920s. In 1926 and 1927, there were 24 fatal commercial airline accidents, and in 1929, 61 people were killed in 51 airline accidents. The year 1929 still remains the worst year on record, with an accident rate of approximately one per million miles flown [49,50]. The numbers of fatalities due to commercial airline accidents in the United States for the period of 1983–1995 are presented in Table 7.2 [49].

For the years 1989–1995, the accident rate per million flight departures in the United States was 0.37, 0.29, 0.33, 0.22, 0.28, 0.27, and 0.40, respectively [49].

It is to be noted that in comparison to other areas, the airline-associated fatalities in the United States are extremely low. For example, in 1995, people were approximately 300 times more likely to die in a motor vehicle accident and approximately 30 times more likely to drown than get killed in an airplane-associated accident [49].

TABLE 7.2

The Number of Fatalities due to
Commercial Airline Accidents in the
United States for the Period of 1983–1995

Year	Number of Fatalities
1983	8
1984	0
1985	486
1986	0
1987	212
1988	255
1989	259
1990	8
1991	40
1992	35
1993	0
1994	228
1995	152

Experiences over the years indicate that there are many causes for the occurrence of airplane crashes. For example, a study of 19 major US domestic jet crashes (defined as one in which at least 10% of the airplane passengers die) occurring during the period 1975–1994 highlighted eight main causes of the occurrence of these crashes [49,51]. These eight causes along with their corresponding number of crashes (in parentheses) are as follows [49,51]:

- Thunderstorm wind shear (4)
- Ground or air collisions (3)
- Ice buildup (3)
- Engine loss (2)
- Hydraulic failure (2)
- Taking off without the flaps in the right position (2)
- Sabotage (1)
- Cause unknown (2)

7.18 Marine Accidents

Over the years, many marine accidents have occurred around the globe. Two of the more noteworthy accidents are described in the following, separately.

7.18.1 The *Estonia* Accident

The *Estonia* accident is concerned with an Estonian-flagged roll-on-roll passenger ferry called *Estonia*, and it occurred on September 28, 1994. On September 27, 1994, *Estonia* left Tallinn, the capital city of Estonia, for Stockholm, Sweden, with 989 passengers on board, and on September 28, 1994, it sank in the Baltic Sea [52]. The accident resulted in 852 fatalities.

A subsequent investigation into the disaster revealed that the bow visor locks of *Estonia* were too weak because of their poor design and manufacture. All in all, during bad weather, the locks broke and the visor fell off by pulling open the inner bow ramp that caused the disaster [52,53].

7.18.2 The *Herald of Free Enterprise* Accident

The *Herald of Free Enterprise* accident is concerned with a passenger ship called *Herald of Free Enterprise*, and it occurred on March 6, 1987. On March 6, 1987, *Herald of Free Enterprise* departed from Zeebrugge Harbor, Belgium, and about 5 minutes after its departure, it capsized. The accident resulted in over 180 fatalities [52,54]. The capsizing of the ship was due to a combination of adverse factors such as the bow door being left open, the vessel speed, and the trim by the bow.

The subsequent public investigation into the disaster is considered as an important milestone in the history of ship safety in the United Kingdom. It resulted in actions such as the following [52]:

- The development of a formal safety assessment process in the shipping industry
- The introduction of the International Safety Management code for the safe operation of ships and for pollution prevention
- Changes to marine safety rules and regulations

7.19 Ship Port-Associated Hazards

There are many ship port-associated hazards, and they may be categorized under eight classifications. These classifications are as follows [55]:

- *Classification 1: Pollution*—This classification contains those hazards that are concerned with the release of material that can cause damage to the surrounding environment. An example of such hazards is crude oil spills.

- *Classification 2: Fire/Explosion*—This classification contains those hazards that are concerned with fire or explosion on the vessel or in the cargo bay. Two examples of such hazards are cargo tank fire/explosion and fire in the engine room.

- *Classification 3: Loss of containment*—This classification contains those hazards that are concerned with the release and the dispersion of dangerous substances. Two examples of such hazards are release of flammables and release of toxic material.

- *Classification 4: Maneuvering*—This classification contains those hazards that are concerned with the failure to position the vessel as intended or to keep the vessel on the right track. Two examples of such hazards are fine-maneuvering error and berthing/unberthing error.

- *Classification 5: Environmental*—This classification contains those hazards that occur when weather exceeds harbor operation criteria or vessel design criteria. Some examples of such hazards are extreme weather, strong currents, and winds exceeding port criteria.

- *Classification 6: Ship related*—This classification contains those hazards that are concerned with ship-specific operations or equipment. Some examples of such hazards are loading/overloading, flooding, mooring failure, and anchoring failure.

- *Classification 7: Impact and collision*—This classification contains those hazards that are concerned with an interaction with a moving or a stationary object or a collision with a vessel. Some examples of such hazards are berthing impacts, vessel collision, and striking while at berth.

- *Classification 8: Navigation*—This classification contains those hazards that have potential for a deviation of the ship from its intended designated route or channel. Some examples of these hazards are navigation error, vessel not under command, and pilot error.

PROBLEMS

1. What are the ways in which software can cause/contribute to a hazard?
2. Describe the role of software quality assurance organization with respect to software safety.
3. List at least 10 methods that can be used to perform software hazard analysis.
4. Discuss robot hazards and safety-related problems.
5. What are the common robot safety-related features and their functions?

6. Describe the following two robot safeguard methods:
   a. Flashing lights
   b. Intelligent systems
7. List at least six truck safety-related facts and figures.
8. Discuss truck and bus safety-related issues.
9. Describe at least four examples of rail accidents and their causes.
10. Discuss the following two accidents
    a. The *Estonia* accident
    b. The *Herald of Free Enterprise* accident

References

1. Cha, S.S., Management aspect of software safety, *Proceedings of the 8th Annual Conference on Computer Assurance*, 1993, pp. 35–40.
2. *An Interpretation of the Technical Guidance on Safety Standards in the Use, etc., of Industrial Robots*, Japanese Industrial Safety and Health Association, Tokyo, 1985.
3. *American National Standard for Industrial Robots and Robot Systems: Safety Requirements*, ANSI/RIA R15.06-1986, American National Standards Institute, New York, 1986.
4. Freund, P.E.S., Martin, G.T., Speaking about accidents: Ideology of auto safety, *Health*, Vol. 1, No. 2, 1997, pp. 167–182.
5. Murray, C.J.L., Lopez, A.D., The global burden of disease in 1990: Final results and their sensitivity to alternative epidemiological perspectives, discount rates, age-weights, disability weights, in *Global Burden of Disease*, edited by C.J.L. Murray and A.D. Lopez, Harvard University Press, Cambridge, MA, 1996, pp. 15–24.
6. Friedman, M.A., Voas, J.M., *Software Assessment*, Wiley, New York, 1995.
7. Leveson, N.G., Software safety: Why, what, and how? *Computing Surveys*, Vol. 18, No. 2, 1986, pp. 125–163.
8. Dhillon, B.S., *Engineering Safety: Fundamentals, Safety, and Applications*, World Scientific Publishing, River Edge, NJ, 2003.
9. Leveson, N.G., Cha, S.S., Shimeall, T.J., Safety verification of ADA programs using software fault trees, *IEEE Software*, Vol. 8, No. 4, 1991, pp. 48–59.
10. Fortier, S.C., Michael, J.B., A risk-based approach to cost-benefit analysis of software activities, *Proceedings of the 8th Annual Conference on Computer Assurance*, 1993, pp. 53–60.
11. Dhillon, B.S., *Computer System Reliability: Safety and Usability*, CRC Press, Boca Raton, FL, 2013.
12. Herrmann, D.S., *Software Safety and Reliability*, IEEE Computer Society Press, Los Alamitos, CA, 1999.
13. Leveson, N.G., *Software*, Addison-Wesley Publishing, Reading, MA, 1995.

14. Mendis, K.S., Software safety and its relation to software quality assurance, in *Handbook of Software Quality Assurance*, edited by G.G. Schulmeyer and J.I. McManus, Prentice Hall, Upper Saddle River, NJ, 1999, pp. 669–679.
15. Ippolito, L.M., Wallace, D.R., *A Study on Hazard Analysis in High Integrity Software Standards and Guidelines*, Report No. NISTIR 5589, National Institute of Standards and Technology, US Department of Commerce, Washington, DC, January 1998.
16. Sheriff, Y.S., Software safety analysis: The characteristics of efficient technical walk-throughs, *Microelectronics and Reliability*, Vol. 32, No. 3, 1992, pp. 407–414.
17. Hansen, M.D., Survey of available software-safety analysis techniques, *Proceedings of the Annual Reliability and Maintainability Symposium*, 1989, pp. 46–49.
18. Hammer, W., Price, D., *Occupational Safety Management and Engineering*, Prentice Hall, Upper Saddle River, NJ, 2001.
19. Ericson, C.A., *Hazard Analysis Techniques for System Safety*, Wiley, New York, 2005.
20. Weik, M.H., *Computer Science and Communications Dictionary*: Vol. 2, Kluwer Academic Publishers, Norwell, MA, 2000.
21. Ziskovsky, J.P., Risk analysis and the R^3 factor, *Proceedings of the Robots 8th Conference*, Vol. 2, June 1984, pp. 15.9–15.21.
22. Ziskovsky, J.P., Working safely with industrial robots, *Plant Engineering*, May 1984, pp. 81–85.
23. Addison, J.H., *Robotic Safety System and Methods: Savannah River Site*, Report No. DPST-84-907 (DE 35-008261), E.I. du Pont de Nemours and Company, Savannah River, Laboratory, Aiken, SC, December 1984.
24. Dhillon, B.S., *Robot System Reliability and Safety: A Modern Approach*, CRC Press, Boca Raton, FL, 2015.
25. Dhillon, B.S., *Robot Reliability and Safety*, Springer-Verlag, New York, 1991.
26. Akeel, H.A., Intrinsic robot safety, in *Working Safely with Industrial Robots*, edited by P.M., Strubhar, Robotics International of the Society of Manufacturing Engineers, Publications Development Department, Dearborn, MI, 1986, pp. 61–68.
27. Clark, D.R., Lehto, M.R., Reliability, maintenance, and safety of robots, in *Handbook of Industrial Robotics*, edited by S.Y. Nof, Wiley, New York, 1999, pp. 717–753.
28. Bararett, R.J., Bell, R., Hudson, P.H., Planning for robot installation and maintenance: A safety framework, *Proceedings of the 4th British Robot Association Annual Conference*, 1981, pp. 18–21.
29. Cox, J., Tips on truck transportation, *American Papermaker*, Vol. 59, No. 3, 1996, pp. 50–53.
30. *The Domain of Truck and Bus Safety Research*, Transportation Research Circular No. E-C117, Transportation Research Board, Washington, DC, 2007.
31. Sheiff, H.E., Status report on large-truck safety, *Transportation Quarterly*, Vol. 44, No. 1, 1990, pp. 37–50.
32. *Large Truck Crash Facts 2003*, Report No. FMCSA-RI-04-033, Federal Motor Carrier Safety Administration, Washington, DC, 2005.
33. Hamilton, S., The top truck and bus safety issues, *Public Roads*, Vol. 59, No. 1, 1995, p. 20.
34. Dhillon, B.S., *Safety and Human Error in Engineering Systems*, CRC Press, Boca Raton, FL, 2013.

35. Saccomanno, F.F., Craig, L., Shortreed, J.H., Truck safety issues and recommendations: Results of the conference on track safety: Perceptions and reality, *Canadian Journal of Civil Engineers*, Vol. 24, 1997, pp. 326–332.
36. Gillen, M., Baltz, D., Gassel, M., Kirsch, L., Vacarro, D., Perceived safety climate, job demands, and coworker support among union and non-union injured construction workers, *Journal of Safety Research*, Vol. 33, 2002, pp. 33–51.
37. *Macdona Accident*, Report No. 04-03, National Transportation Safety Board, Washington, DC, 2004.
38. *Derailment of Amtrak Auto Train P052-18 on the CSXT Railroad Near Crescent City, FL*, Report No. RAR03/02, National Transportation Safety Board, Washington, DC, 2003.
39. Transportation Accident Investigation Commission, Wellington. Available online at www.taic.org.nz (retrieved on July 26, 2011).
40. MacAongusa, B., *The Harcourt Street Line: Back on Track*, Currach Press, Dublin, 2003.
41. Murray, D., Collision at Dundrum, *Journal of the Irish Railway Record Society*, Vol. 17, No. 116, 1991, pp. 434–441.
42. MacAongusa, B., *Broken Rails*, Currach Press, Dublin, 2005.
43. Schneider, W., Mase, A., *Railway Accidents of Great Britain and Europe: Their Causes and Consequences*, David and Charles, Newton Abbot, 1970.
44. Rolt, L.T.C., *Red for Danger*, David and Charles, Newton Abbot, 1966.
45. Wikipedia, Classification of railway accidents, retrieved on December 10, 2010. Available online at www.en.wikipedia.org/wiki/classification-of-railway-accidents, 2010.
46. *TR News*, Ensuring railroad tank car safety, No. 176, January–February 1995, pp. 30–31.
47. Wells, A.T., Rodrigues, C.C., *Commercial Aviation Safety*, McGraw Hill Book Company, New York, 2004.
48. *Statistical Summary of Commercial Jet Aircraft Accidents: Worldwide Operations 1959–2001*, Boeing Commercial Airplane Company, Seattle, WA, 2001.
49. Benowski, K., *Safety in the Skies*, Quality Progress, January 1997, pp. 25–35.
50. Aviation accident synopses and statistics, National Transportation Safety Board, Washington, DC, 2007. Available online at www.ntbs.gov/Aviation/Aviation.htm.
51. Beck, M., Hosenball, M., Hager, M., Springen, K., Rogers, P., Underwood, A., Glick, D., Stanger, T., How safe is this flight? *Newsweek*, 1995, April 24, pp. 18–29.
52. Wang, J., Maritime risk assessment and its current status, *Quality and Reliability Engineering International*, Vol. 22, 2006, pp. 3–19.
53. Wang, J., A brief review of marine and offshore safety assessment, *Marine Technology*, Vol. 39, No. 2, 2002, pp. 77–85.
54. *Herald of Free Enterprise: Fatal Accident Investigation*, Report No. 8074, United Kingdom Department for Transport, Her Majesty's Stationery Office, London, 1987.
55. Trbojevis, V.M., Carr, B.J., Risk based methodology for safety improvement in ports, *Journal of Hazardous Materials*, Vol. 71, 2000, pp. 467–480.

8

Medical and Mining System Safety

8.1 Introduction

Each year, billions of dollars are spent around the globe to produce various types of medical systems/devices for use in the area of healthcare. A medical system/device must not only be reliable but also be safe for users and patients.

In regard to health and safety in the United States, the passage of the Occupational Safety and Health Act by the US Congress in 1970 is considered to be an important milestone. Other important milestones that are specifically concerned with medical devices in the United States are the Medical Device Amendments of 1976 and the Safe Medical Device Act in 1990.

Nowadays, a vast sum of money is spent each year to produce various types of mining equipment/systems around the globe. Over the years, many accidents have occurred involving various types of mining equipment/ systems. In order to improve safety in the US mine (including equipment/ systems), in 1970, the US Congress passed the Mine Safety and Health Act. As a result of this act, the US Department of Labor established an agency called the Mine Safety and Health Administration (MSHA). The main goal of MSHA includes items such as promotion of better health and safety conditions in the mines, reduction of health-associated hazards, and enforcement of compliance with mine safety and health standards [1].

This chapter presents various important aspects of medical and mining system safety.

8.2 Medical System Safety-Related Facts and Figures

Some of the facts and figures, directly or indirectly, concerned with medical system safety are as follows:

- As per the study by Dhillon [2], the Emergency Care Research Institute (ECRI) after examining a sample of 15,000 hospital products concluded that 4–6% of these products were dangerous enough to warrant immediate corrective action.
- As per the study of Schneider and Hines [3], faulty software programs in heart pacemakers resulted in two deaths.
- In 1969, the special committee of the US Department of Health, Education, and Welfare reported that over a period of 10 years, there were approximately 10,000 medical device-related injuries and 731 resulted in fatalities [4,5].
- As per the study by the ECRI [6], a 5-year-old patient was crushed to death beneath a pedestal-style electric bed in which the child was placed after hospital admission.
- As per the study by Casey [7], a patient fatality occurred because of radiation overdose involving a Therac radiation therapy device.
- As per the ECRI [8], over time, ECRI has received a large number of reports concerning radiologic equipment-related failures that either caused or had the potential for causing serious patient injury or death.

8.3 Safety-Related Requirements for Medical Devices/Systems and Types of Medical Device/System Safety

Over the years, government and other agencies have placed various types of, directly or indirectly, safety-related requirements on medical devices. These requirements may be categorized under the following three areas [9]:

- *Area 1: Safe design*—The requirements of area 1 are mechanical hazard prevention, care for environmental conditions, excessive heating prevention, care for hygienic factors, protection against radiation hazards, protection against electrical shock, and proper material in regard to biological, mechanical, and chemical factors. Mechanical hazard prevention includes factors such as device stability, breaking strength, and safe distances.

The care for environmental conditions includes factors such as humidity, electromagnetic interactions, and temperature. The excessive heating prevention-associated mechanisms are effective design, cooling, and temperature control. All the remaining requirements are considered self-explanatory; however, the additional information on them is available in the study by Leitgeb [9].

- *Area 2: Safe function*—The elements of area 2 are the accuracy of measurements, warning for or prevention of hazardous outputs, and reliability.
- *Area 3: Sufficient information*—Area 3 includes items such as instructions for use, labeling, packaging, and accompanying documentation.

Medical device/system safety may be classified under three types, as shown in Figure 8.1 [9].

Unconditional safety is most effective and is preferred over all other types or possibilities. However, conditional safety calls for the eradication of all device/system-associated risks through design. Furthermore, it is to be noted that the employment of warnings certainly complements satisfactory device/system design but does not replace it.

Conditional safety is used in circumstances when unconditional safety cannot be realized. For example, in the case of an X-ray/laser device, it is completely impossible to avoid dangerous radiation emissions. Nonetheless, it is well within the means for minimizing risk with measures, such as incorporating a locking mechanism that permits device activation by authorized personnel only or limiting access to therapy rooms. Examples of the indirect safety means are X-ray folding screens and protective laser glasses.

Finally, the descriptive safety is employed in circumstances when it is not possible or inappropriate for providing safety through the two means previously mentioned (i.e., unconditional or conditional). However, descriptive safety with respect to operation, maintenance, mounting, transport, connection, and replacement may simply be statements, such as *This Side Up*, *Handle with Care*, and *Not for Explosive Zones*.

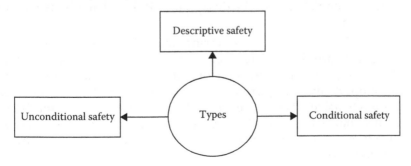

FIGURE 8.1
Types of medical device/system safety.

8.4 Safety in Medical Device/System Life Cycle

Experiences over the years clearly indicate that in order to have safe medical devices/systems, safety has to be considered throughout their entire life cycle. Thus, the life cycle of a medical device/system may be divided into five phases as follows [10–12]:

1. *Phase 1: Concept phase*—In Phase 1, past data and future technology-associated projections become the basis for the device/system under consideration and safety-related problems are highlighted and examined. The preliminary hazard analysis (PHA) method is a very useful tool for highlighting hazards during this phase. At the end of this phase, some of the typical questions to ask with respect to device/system safety are as follows [10–12]:

   - Are all the hazards highlighted and appropriately examined for developing hazard controls?
   - Is the risk analysis initiated for developing mechanisms for hazard controls?
   - Are all the fundamental safety design-associated requirements properly in place so that the definition phase can be started?

2. *Phase 2: Definition phase*—The main objective of phase 2 is to provide proper verification of the initial design and engineering concerned with the medical device/system under consideration. The results of the PHA are updated along with the initiation of subsystem hazard analysis as well as their ultimate integration into the overall device/system hazard analysis. Methods, such as fault tree analysis (FTA) and fault hazard analysis, may be used for examining certain known hazards as well as their effects.

   All in all, the system definition will initially lead to the acceptability of a desirable general device/system design even though, because of the incompleteness of the design, not all the related hazards will be known.

3. *Phase 3: Development phase*—During phase 3 of the device/system, the efforts are directed on areas such as operational use, producibility engineering, environmental impact, and integrated logistic support. By using prototype analysis and testing results, the comprehensive PHA is performed to evaluate human–machine-related hazards, in addition to developing PHA further because of more completeness of the design of the device/system under consideration.

4. *Phase 4: Production phase*—In phase 4, the device/system safety engineering report is prepared with the aid of data collected

during the phase. The report documents and highlights the device/system-related hazards.

5. *Phase 5: Deployment phase*—In phase 5, the data concerning failures, accidents, incidents, etc., are collected, and safety professionals review any changes to the device/system. The device/system safety analysis is updated as appropriate.

8.5 Classifications of Medical Device/System Accident Causes and Methods for Conducting Medical Device/System Safety Analysis and Considerations for Their Selection

Over the years, it has been observed that there are many causes for the occurrence of medical device/system-related accidents. The professionals working in the area have classified these causes under seven classifications as shown in Figure 8.2 [13].

Additional information on the classifications shown in Figure 8.2 is available in the study by Brueley [13].

There are many methods that can be used for conducting safety analysis of medical devices/systems. Some of these methods are as follows [10,11,14–17]:

• Operating hazard analysis (OHA)
• FTA

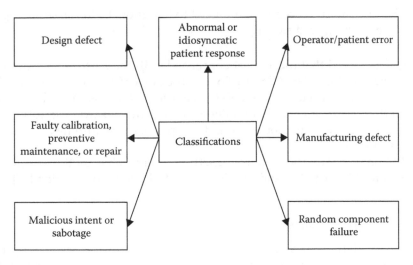

FIGURE 8.2
Classifications of medical device/system accident causes.

- Human error analysis
- Interface safety analysis (ISA)
- Failure modes and effect analysis (FMEA)
- Technique of operation review

The first three of the previously mentioned six methods are presented in the following sections, and the remaining three methods (i.e., ISA, FMEA, and technique of operation review) are described in Chapter 4.

8.5.1 Operating Hazard Analysis

OHA is a method that particularly focuses on hazards occurring from tasks/activities for operating system functions that take place as the system is used, transported, or stored. Generally, the OHA is initiated early in the system development cycle so that proper inputs to technical orders are provided, which in turn govern the system testing. The application of the OHA provides a basis for safety considerations, such as follows [10,11,12,15]:

- Highlighting of system/item functions relating to hazardous occurrences
- Development of emergency procedures, warning, or special instructions with respect to operation
- Design modifications for eradicating hazards
- Safety guards and safety devices
- Special safety procedures in regard to handling, training, servicing, transporting, and storing

It is to be noted that the analyst concerned with the performance of OHA needs engineering-related descriptions of the device/system under consideration with available support facilities. In addition, OHA is carried out using a form that needs information on items such as the hazard description, the hazard effects, the operational event description, the requirements, and the hazard control.

Additional information on this method is available in the studies by Roland and Moriarty [11] and the Electronic Industries Association [15].

8.5.2 Fault Tree Analysis

FTA is a widely used method in the industrial sector for performing reliability analysis of engineering systems. The method was developed in the early 1960s at the Bell Telephone Laboratories to perform the analysis of the

Minuteman Launch Control System [18]. Some of the main points concerned with this method are as follows [10,12,18]:

- It allows users to evaluate alternatives and pass appropriate judgment on acceptable trade-offs among them.
- It is a quite useful tool to analyze operational devices/systems for undesirable or desirable occurrences.
- It is a very useful analysis in the early design phases of new systems/devices.
- It can be used for evaluating certain operational functions (e.g., start-up or shutdown phases of facility/system/device operation).

Additional information on FTA is available in Chapter 4 and in the studies by Dhillon [10] and Dhillon and Singh [18].

8.5.3 Human Error Analysis

Human error analysis is considered quite useful for highlighting hazards prior to their occurrence in the form of accidents. There could be the following time approaches to human error analysis:

1. Observing personnel/workers during their work hours with respect to hazards
2. Performing tasks for obtaining firsthand information on hazards

All in all, irrespective of the performance of the human error analysis, it is strongly recommended to perform it in conjunction with hazard and operability analysis and FMEA methods described in Chapter 4.

Additional information on human error analysis is available in the studies by Roland and Moriarty [11], Goetsch [14], the Electronic Industries Association [15], Hammer [16], and Gloss and Wardle [17].

8.5.4 Considerations for the Selection of Safety Analysis Methods for Conducting Medical Device/System Safety Analysis

Experiences over the years clearly indicate that conducting effective safety analysis of medical systems/devices requires a careful consideration in the selection and the implementation of proper safety analysis methods for given situations. Thus, questions such as those presented in the following should be asked prior to the selection and the implementation of safety analysis methods for the situation under consideration [10,15].

- What type of data, information, etc., is required prior to the initiation of the study?
- When are the results needed?

- Who are the users of the results?
- What is the time frame for the start of analysis as well as for its completion, submission, review, and update?
- What mechanism is needed for acquiring information from subcontractors (if applicable)?

8.6 Mining Equipment/System Safety-Related Facts and Figures and Injuries and Fatalities due to Crane, Drill Rig, and Haul Truck Contact with High-Tension Power Lines

Some of the facts and figures, directly or indirectly, concerned with mining equipment safety are as follows:

- As per the study by Cawley [19], during the period of 1990–1999, electricity was the fourth leading cause for the occurrence of fatalities in the US mining industry.
- In 2004, approximately 17% of the injuries that occurred in the US underground coal mines were connected to bolting machine equipment [20].
- A study conducted by the US Bureau of Mines (now the National Institute for Occupational Safety and Health [NIOSH]) reported that equipment was the main cause of injury in approximately 11% of all mining accidents and a secondary causal factor in another 10% of the accidents [21–23].
- As per the study by Rethi and Barett [24], during the period of 1983–1990, approximately 20% of the coal mine-related injuries occurred during equipment/system maintenance or while using various types of hand tools.
- As per the study by the MSHA [25], during the period of 1978–1988, maintenance-related activities in the US mines accounted for approximately 34% of all lost time injuries.
- As per the study by De Rosa [26], during the period of 1990–1999, 76 injuries were due to 197 equipment fires in coal mining operations in the United Sates.

High-tension or overhead electric power lines present a quite serious electrocution hazard to individuals working in various industries, because equipment such as drill rigs, haul trucks, and cranes is frequently exposed to these lines. When contacting the power lines, this type of equipment/system becomes elevated to a very high voltage, and simultaneous contact to the *hot* frame and ground by people can lead to dangerous electric shocks and burns.

The mining industry is one of those industries where the occurrence of such accidents is greatest. In the United States, each year, about 2300 accidental overhead power line contacts occur [27]. For the period of 1980–1997, the US mining industrial sector reported at least 94 mobile equipment overhead line contact-related accidents [27]. These accidents caused 114 injuries, and about 33% of them caused fatalities. Most of these accidents involved drills (14%), dump bed trucks (24%), and cranes (47%).

8.7 Human Factors-Related Tips for Safer Mining Equipment/Systems

As human factors play an important role in the safety of mining equipment/systems, some of the human factors-related tips for safer mining equipment/systems are as follows [28,29]:

- Ensure that seats are designed in such a way that miners can maintain or replace it with ease.
- Ensure that the relative placement of controls and displays for similar equipment/systems/machines is maintained effectively.
- Ensure that all involved operators can identify all the necessary controls quickly and accurately.
- Anticipate potential safety-related hazards and appropriate emergency measures prior to starting the design process.
- Ensure that the workstation provides an unobstructed line of site to locations/objects that should be clearly visible for performing a stated task safely.
- Aim to allocate workloads as evenly as possible between hands and feet.
- Ensure that each and every design control can effectively guard/withstand against possible abuse (e.g., from falling roofs or from the forces imposed during a panic response in emergencies).
- Ensure that there is a sufficient contrast between the object luminance or the location of interest and the surrounding background, so that a specified task can be carried out safely.
- Ensure that the workstation effectively fits all potential operators from the 5th to the 95th percentile range.
- Ensure that the seats do not hinder an operator's ability to control the machine or equipment/system.

- Ensure that the seats provide appropriate design features for guarding against shocks due to minor collisions or rough roads that could tend to unseat the involved individual.
- Ensure that the seats adjust and properly fit to body dimensions, distribute weight for relieving pressure points, and support posture.
- Ensure that each and every design control appropriately complies with anthropometry-related data concerning human operators.
- Ensure that the seats do not hinder the operator's ability when exiting or entering the workstation.

8.8 Causes of Mining Equipment-Related Accidents and Mining Equipment Maintenance-Related Accidents

Over the years, many studies have been performed to highlight the occurrence of mining equipment-related accidents. One such study conducted by the US Bureau of Mines (now NIOSH) has highlighted the seven causes shown in Figure 8.3 for the occurrence of mining equipment-related accidents [29,30].

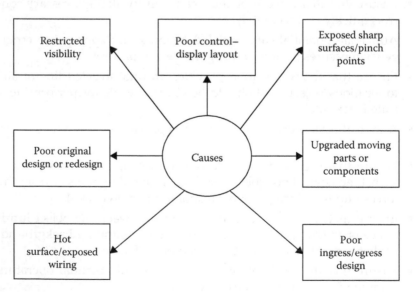

FIGURE 8.3
Causes for the occurrence of mining equipment-related accidents.

During the mining equipment maintenance activity, there are many types of accidents that can occur. Some of the commonly occurring such accidents are as follows [31]:

- Contact with hot objects
- Impact with an object or machinery
- Falling on objects
- Overexertion
- Falling objects
- Inhalation of noxious fumes
- Push–pull
- Flying objects

Additional information on the previously mentioned accidents is available in the study by Unger and Conway [31].

8.9 Methods for Performing Mining Equipment/ System Safety Analysis

There are many methods that can be used to perform mining equipment/ system safety analysis. Three of these methods are presented in the following sections.

8.9.1 Management Oversight and Risk Tree Analysis

Management oversight and risk tree (MORT) analysis is an effective safety assessment method that can basically be applied to any safety-related program in the area of mining, and it first appeared in 1973 [32]. The method particularly focuses on administrative/programmatic control of hazardous conditions and is specifically designed for identifying, evaluating, and preventing safety-related oversights, omissions, and errors by workers and management that can lead to accidents.

The method is composed of nine steps presented in the following [32,33]:

- *Step 1:* Obtain proper working knowledge of the equipment/system under consideration.
- *Step 2:* Choose the accident to be analyzed.
- *Step 3:* Highlight all possible hazardous energy flows and barriers related to the sequence of accident.

- *Step 4:* Document required information in the standard MORT-type analytical tree format.
- *Step 5:* Determine all possible factors that cause initial unwanted energy flow.
- *Step 6:* Document all the safety program elements that are considered to be less than adequate with respect to the unwanted energy flow.
- *Step 7:* Continue conducting analysis of the safety program-related elements with respect to the rest of the unwanted energy flows (if any).
- *Step 8:* Determine all the management system factors related to the potential accident.
- *Step 9:* Evaluate the accomplished analysis for all safety program-related elements that could be useful in lowering the likelihood of the potential accident occurrence.

Some of the main advantages of this method are as follows [32,33]:

- The findings of MORT analysis can suggest necessary improvements to an ongoing safety program that could be very helpful to reduce property damage and injuries and save lives.
- Useful for examining management, hardware, and human aspects of an industrial system/equipment because they collectively cause accidents.
- Comprehensive and effective approach that attempts to evaluate all aspects of safety in any work activity.

In contrast, some of the main disadvantages of this method are as follows [32,33]:

- Emphasizes management's responsibility to provide a safe work environment
- Is a time-consuming approach
- Creates a large amount of complex detail

8.9.2 Binary Matrices

Binary matrix analysis is a quite useful, logical, and qualitative method for identifying system interactions [34]. It can be used during the system description stage of safety analysis or as a final checkpoint in a PHA or a FMEA, to ensure that each important dependency related to the system under study has been considered in the analysis properly.

The binary matrix is the specific tool used in binary matrices, and it contains information concerning the relationships between the system elements. The main objective of this matrix is to highlight the one-on-one dependencies between all elements of a system under study. All in all, the matrix serves as a very useful tool to *remind* the involved analyst that failures in one part of a given system may affect the normal operation of the other subsystems in completely distinct areas.

The application of this method in the area of mining is demonstrated in Daling and Geffen [35].

8.9.3 Consequence Analysis

Consequence analysis is a method that is concerned with determining the impact of the occurrence of an undesired event on items such as adjacent property, people, or environment. Some examples of such events are fire, explosions, the release of toxic materials, or projection of debris. The basic consequences of concern in the mining area include injuries, deaths, and losses due to equipment/system/property damage and operational downtime.

Needless to say, consequence analysis generally serves as one of the intermediate steps of safety analysis, as the consequences of an accident are generally determined initially using methods such as FMEA or PHA. Additional information on the application of this method in the mining industrial sector is available in the study by Daling and Geffen [35].

PROBLEMS

1. List at least five medical system safety-related facts and figures.
2. What are the medical devices/system safety-related requirements?
3. Discuss safety in the medical device/system life cycle.
4. What are the classifications of medical device/system accident causes?
5. Describe OHA.
6. List at least six methods that can be used to perform medical device/system safety analysis. Also, discuss the considerations for the selection of these methods.
7. List at least five mining equipment/system safety-related facts and figures.
8. List at least 10 of the most useful human factors-related tips for safer mining equipment/systems.
9. What are the causes for the occurrence of mining equipment accidents?
10. Describe MORT analysis.

References

1. Mine Safety and Health Administration (MSHA), Department of Labor, Washington, DC. Available from www.msha.gov.
2. Dhillon, B.S., Reliability technology in health care systems, *Proceedings of the IASTED International Symposium on Computers*, Advanced Technology in Medicine, and Health Care Bioengineering, 1990, pp. 84–87.
3. Schneider, P., Hines, M.L.A., Classification of medical software, *Proceedings of IEEE Symposium on Applied Computing*, 1990, pp. 20–27.
4. *Medical Devices: Hearings before the Subcommittee on Public Health and Environment*, US Congress Interstate and Foreign Commerce, Serial No. 93-61, US Government Printing Office, Washington, DC, 1973.
5. Banta, H.D., The regulation of medical devices, *Preventive Medicine*, Vol. 19, 1990, pp. 693–699.
6. Emergency Care Research Institute (ECRI), *Electric Beds Can Kill Children*, Medical Device Safety Report, Emergency Care Research Institute, Plymouth Meeting, PA, 2001.
7. Casey, S., Set phasers on stun: And other true tales of design technology and human error, Agean, Inc., Santa Barbara, CA, 1993.
8. ECRI, *Mechanical Malfunctions and Inadequate Maintenance of Radiological Devices*, Medical Device Safety Report, Emergency Care Research Institute, Plymouth Meeting, PA, 2001.
9. Leitgeb, N., *Safety in Electromedical Technology*, Interpharm Press, Buffalo Grove, IL, 1996.
10. Dhillon, B.S., *Medical Device Reliability and Associated Areas*, CRC Press, Boca Raton, FL, 2000.
11. Roland, H.E., Moriarty, B., *System Safety Engineering and Management*, Wiley, New York, 1983.
12. Dhillon, B.S., *Safety and Human Error in Engineering Systems*, CRC Press, Boca Raton, FL, 2013.
13. Brueley, M.E., Ergonomics and error: Who is responsible? *Proceedings of the First Symposium on Human Factors in Medical Devices*, 1989, pp. 6–10.
14. Goetsch, O.L., *Occupational Safety and Health*, Prentice Hall, Englewood Cliffs, NJ, 1996.
15. *System Safety Analytical Techniques*, Safety Engineering Bulletin, Electronic Industries Association, Washington, DC, No. 3, May 1971.
16. Hammer, W., *Product Safety Management and Engineering*, Prentice Hall, Englewood Cliffs, NJ, 1980.
17. Gloss, D.S., Wardle, M.G., *Introduction to Safety Engineering*, Wiley, New York, 1984.
18. Dhillon, B.S., Singh, C., *Engineering Reliability: New Techniques and Applications*, Wiley, New York, 1981.
19. Cawley, J.C., Electrical accidents in the mining industry, 1990–1999, *IEEE Transactions on Industry Applications*, Vol. 39, No. 6, 2003, pp. 1570–1576.
20. Burgess-Limerick, R., Steiner, L., Preventing injuries: Analysis of injuries highlights high priority hazards associated with underground coal mining equipment, *American Longwall Magazine*, 2006, August, pp. 19–20.

21. Unger, R.L., Tips for safer mining equipment, US Department of Energy's Mining Health and Safety Update, Vol. 1, No. 2, August 1996, pp. 14–15.
22. What causes equipment accidents? National Institute for Occupational Safety and Health, Washington, DC, 2008. Available from www.ede.gov/niosh /mining/topics/machinesafety/equipment%20dsgn/equipmentaccident.
23. Sanders, M.S., Shaw, B.E., *Research to Determine the Contribution of System Factors in the Occurrence of Underground Injury Accidents*, Report No. USBM OFR 26-89, US Bureau of Mines (USBM), Washington, DC, 1988.
24. Rethi, L.L., Barett, E.A., *A Summary of Injury Data for Independent Contractor Employees in the Mining Industry from 1983–1990*, Report No. USBMIC 9344, US Bureau of Mines, Washington, DC, 1990.
25. Mine Safety and Health Administration (MSHA), *MSHA Data for 1978–1988*, Mine Safety and Health Administration, US Department of Labor, Washington, DC.
26. De Rosa, M., Equipment fires cause injuries: Recent NIOSH study reveals trends for equipment fires at US coal mines, *Coal Age*, 2004, October, pp. 28–31.
27. Sacks, H.K., Cowley, J.C., Homce, G.T., Yenchek, M.R., Feasibility study to reduce injuries and fatalities caused by contact cranes, drill rigs, and haul trucks with high-tension lines, *IEEE Transactions on Industry Applications*, Vol. 37, No. 3, 2001, pp. 914–919.
28. Unger, R.L., Tips for safer mining equipment, *Holmes Safety Association Bulletin*, 1996, October, pp. 3–4.
29. Dhillon, B.S., *Mine Safety: A Modern Approach*, Springer, Inc., London, 2010.
30. Sanders, M.S., Shaw, B.E., *Research to Determine the Contribution of System Factors in the Occurrence of Underground Injury Accidents*, Report No. USBM OFR 26-89, US Bureau of Mines, Washington, DC, 1988.
31. Unger, R.L., Conway, K., Impact of maintainability design on injury rates and maintenance costs or underground mining equipment, in *Improving Safety at Small Underground Mines*, compiled by R.H. Peters, Report No. 18-94, US Bureau of Mines, Department of the Interior, Washington, DC, 1994.
32. Johnson, W.G., *The Management Oversight and Risk Tree-MORT*, Report No. SAN 821-1, US Atomic Energy Commission, Washington, DC, 1973.
33. Dhillon, B.S., *Mining Equipment Reliability, Maintainability, and Safety*, Springer, Inc., London, 2008.
34. Cybulskis, P. et al., *Review of Systems Interaction Methodologies*, Report No. NUREG/CR-1896, Battelle Columbus Laboratories, Columbus, OH, 1981.
35. Daling, P.M., Geffen, C.A., *User's Manual of Safety Assessment Methods for Mine Safety Officials*, Report No. BuMines OFR 195-(2)-83, US Bureau of Mines, Department of the Interior, Washington, DC, 1983.

9

Software Maintenance and Reliability-Centered Maintenance

9.1 Introduction

Today, computers find applications in virtually all areas of life, and each year, a vast sum of money is spent to develop various types of computer software around the globe. Software maintenance is an important element of the software life cycle, and it may simply be defined as the process of modifying the system/component subsequent to delivery to rectify faults, improve performance or other attributes, or adapt to a change in the use environment [1,2].

Nowadays, annually, organizations around the globe spend a vast sum of money on software maintenance. For example, in 1983, the US Department of Defense (DOD) alone spent US$2 billion on software maintenance, while in the mid-1980s, the figure for the entire United States was around US$30 billion per year [3,4].

Reliability-centered maintenance (RCM) is a systematic process to determine what has to be accomplished for ensuring that any physical facility is able to continuously satisfy its designed functions in its current operating context. Any organization can benefit from RCM if its breakdowns account for greater than 20–25% of the total maintenance workload, as RCM leads to a maintenance program that focuses preventive maintenance (PM) on specific failure modes likely to occur [5].

The term *reliability-centered maintenance* appeared for the first time as the title of a report on the processes employed by the civil aviation industry for preparing programs for aircraft [6,7]. The report, which was prepared by United Airlines, was commissioned by the US DOD in 1974 [8]. Additional information on the history of RCM is available in the studies by Moubray [6], August [7], Nowlan and Heap [8], Smith [9], the National Aeronautics and Space Administration [10], and Dhillon [11].

This chapter presents various important aspects of software maintenance and RCM.

9.2 Software Maintenance-Related Facts and Figures

Some of the facts and the figures, directly or indirectly, concerned with software maintenance are as follows:

- As per the study by Horowitz [12], in the early 1990s, the US DOD spent about US$30 billion annually on software, and approximately two-thirds of that amount was devoted to sustaining deployed software systems.
- As per the study by Foster [13], software maintenance accounts for between 40% and 90% of total life cycle costs.
- A study carried out by the Boeing Company revealed that, on average, approximately 15% of the lines of source code in a simple/easy program are changed each year; 5%, in medium programs; and 1%, in difficult programs [14].
- The US government spends around 40% of the total software-related cost on maintenance [15].
- As per the study by Fairley [16], a software product's typical life span is 1–3 years in development and 5–15 years in use (maintenance).
- A Hewlett-Packard study reported that 60–80% of its research and development personnel are involved in the maintenance of existing software [17].
- As per the studies by Stevenson [18] and Glass [19], most of software maintenance is development in disguise, of which approximately 20% is correction of errors.
- As per the study by Dhillon [4], over two-fifths of software maintenance-related activities are due to modifications and extensions required by the users.
- As per the study by Charette [20], over 80% of a software product's life is spent in maintenance.

9.3 Software Maintenance Problems and Maintenance Types

The maintenance of software systems is quite difficult because they are already operational. Therefore, it is necessary to keep proper balance between the need for change and keeping the system accessible for its users. Many management- and technology-related problems occur when changing software cheaply and quickly. In addition, as many software systems under maintenance are fairly complex and large, the solutions may work quite well for laboratory-scale pilots but fail to scale up to life-sized or industrial software.

Some of the people-related software maintenance problems are low morale of the involved professionals and poor comprehension of the maintenance needs. A clear comprehension of what needs to be changed is very important because approximately 47% of the software maintenance effort is concerned with comprehending the software to be modified [21]. This high figure results from the number of interfaces that require to be examined when changing a component. Furthermore, over 50% of the problems of the involved maintenance professionals are because of the user's poor understanding or skills [22].

An important reason for the low morale is the second-class status frequently accorded to maintenance personnel. The findings of a study indicate that around 12% of the problems during maintenance are due to low morale and productivity [22].

In regard to technology-related maintenance problems, a change made at one place in the software system can have a ripple effect elsewhere. It means that a clear understanding of the consequences of changes is essential. In order for a change to be consistent, maintenance personnel must properly investigate the possibility of occurrence of all types of ripple effects. The propagation of ripple effect may be expressed as a phenomenon by which modifications carried out to a software component during the software life cycle (i.e., test, code, design, or specification phase) affect other components or elements [23].

Software maintenance focuses on the following four aspects of system evolution simultaneously [24]:

- PM (i.e., preventing the degradation of system performance to unacceptable levels)
- Perfective maintenance (i.e., perfecting existing and acceptable functions)
- Adaptive maintenance (i.e., maintaining control over modifications associated with the system)
- Corrective maintenance (i.e., maintaining control over the day-to-day operations of the system)

A survey of 487 software development organizations revealed the percentage distribution of PM, perfective maintenance, adaptive maintenance, and corrective maintenance as 4%, 50%, 25%, and 21%, respectively [22]. Each of the maintenance types is described more clearly in the following.

PM modifies software for enhancing potential reliability/maintainability or provides an improved basis for potential enhancements. Generally, PM is practiced when the involved software professionals find an actual or a potential fault that has not yet become a failure and take appropriate corrective actions. However, this type of maintenance is still relatively rare.

Perfective maintenance modifies existing functions, makes general enhancements, and adds capabilities. It involves carrying out modifications for enhancing some aspects of the system, even when such modifications are not dictated by faults. Improving only modules with a quite high usage, a reasonable life span, and a high cost to perform corrective or adaptive maintenance can be very useful in perfective maintenance [25].

Adaptive maintenance modifies software to effectively interface with a changing environment (i.e., both software and hardware). It is to be noted that the adaptive changes made for adding parameters do not rectify faults; they only allow the system to adapt properly as it evolves. Furthermore, striving for hardware independence and making use of a high-level language improve adaptive maintenance [25].

Finally, corrective maintenance process incorporates the diagnosis and the rectification of errors. For controlling the day-to-day system functions, maintenance professionals respond to problems arising from faults. Some of the ways for improving corrective maintenance are using structured methods, keeping modules as small as possible, and employing high-level languages [25].

9.4 Software Maintenance Methods

Over the years, many methods have been developed that directly or indirectly concern software maintenance. Three of these methods are described in the following sections, separately.

9.4.1 Impact Analysis

Software maintenance depends on and starts with the requirements of a user or a customer. A requirement translating into a seemingly minor change is often more extensive and, consequently, more expensive to implement than originally anticipated. Under such situations, a proper study of the impact of the change could provide very useful information, particularly where change is sophisticated and complex.

Impact analysis may simply be described as the determination of risks associated with the proposed change, including the estimation of effects on factors such as schedule, resources, and effort. A number of ways for measuring the impact of a change are given in the study by Pfleeger and Bohner [26].

9.4.2 Maintenance Reduction

Reduction in the amount of maintenance helps to enhance maintenance productivity. Maintenance personnel armed with the latest knowledge, skills, and methods can reap significant quality and productivity improvements.

The methods that can be used to reduce software maintenance are as follows [4,16,18,19,27–31]:

- Classify the functions under two categories: Inherently more stable and most likely to be changed.
- Encourage good communication among maintenance programmers.
- Usage of portable languages, tools, and operating systems.
- Introduce structured maintenance that makes use of approaches to document currently existing systems and includes guidelines for reading programs, etc.
- Highlight possible enhancements and design the software in such a way that it can easily incorporate those enhancements.
- Schedule maintenance on specific dates only and do not permit any changes in between those dates.
- Use PM methods, such as using limits for tables that are reasonably more than may possibly be required.
- Store constants in tables rather than scattering them throughout the program.
- During software design, consider human factors in areas such as screen layouts. This is one source of frequent modifications/changes.
- Use standard methodologies all the time.

9.4.3 Software Configuration Management

Software configuration management may simply be expressed as a set of tracking and control activities that starts at the beginning of a software development project and stops at the software retirement. Keeping track of the changes made and their effects on other system parts or components is a very challenging task. Generally, the more sophisticated and complex the system, the more components or parts a change will affect. For this reason, configuration management is a very important and critical factor during maintenance.

Configuration management is practiced by forming a configuration control board because many maintenance-related changes are requested by customers/users for correcting failures or making enhancements. The board oversees the entire change process, and its membership includes interested parties: developers, users, and customers. Each and every highlighted problem is handled as follows [24]:

1. A customer, a user, or a developer who finds a problem uses a formal change control form for recording all associated systems. Similarly, in the case of enhancement, all relevant information is documented.
2. The configuration control board members are formally informed of the proposed change.

3. The board members meet and discuss the proposed change.

4. After making a decision regarding the change requested, the board members prioritize the change and assign appropriate individual(s) for making the change.

5. The designated individual(s) highlights the problem source and identifies the changes needed. Working with the test copy, the assigned individual(s) tests and implements the required changes.

6. The designated individual(s) works with the software program librarian for tracking and controlling the modification or for changing the installation in the operational system and updating the related documentation.

7. A change report explaining the changes carried out is filed by the designated individual(s).

It is to be noted that step 6 previously mentioned is the most critical step, because, at any time, the configuration management team members must be fully aware of the status of any component/document in the system. This requires good communication among all involved individuals. Consequently, it is essential to have answers to the following questions [4,32]:

- Who authorized the change?
- Who carried out the change?
- What was actually changed?
- Who can stop the change request?
- When did the change take place?
- Was the change carried out effectively and correctly?
- What is the change priority?
- Who is responsible for the change in question?
- Who was made aware of the change?

The nine questions previously mentioned take into consideration authorization, identification, naming, cancellation, synchronization, authentication, valuation, delegation, and routing, respectively.

9.5 Software Maintenance Costing

Nowadays, the cost of maintaining software systems over their entire life cycles has become a very important issue. For example, in the 1970s, development consumed most of the budget of a software system, and in the 1990s, some estimates indicated that maintenance costs might have increased to as high as 80% of the life cycle cost of a system [24].

In general, the following may be said in regard to software maintenance costs [4]:

- Both nontechnical and technical factors affect maintenance costs quite significantly.
- Software maintenance costs are usually greater than development costs.
- Normally, aging software has high support costs due to old languages, compilers, etc.
- Increase in software is maintained because maintenance corrupts the software structure, thus making further maintenance difficult.

Maintenance costs are affected by many factors as shown in Figure 9.1 [4].

Over the years, many mathematical models, directly or indirectly, concerned with software maintenance costing have been developed. One such model is presented in the following, and information on other models is available in the studies by Shooman [33], Mills [34], and Belady and Lehman [35].

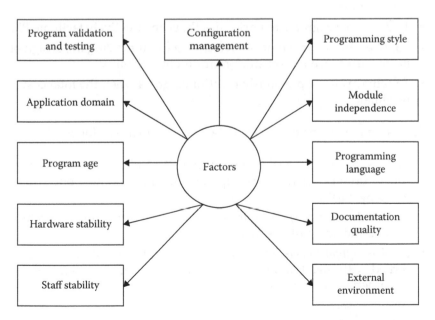

FIGURE 9.1
Factors affecting software maintenance costs.

9.5.1 Maintenance Cost Model

Maintenance cost model is a mathematical model that is considered quite useful for directly estimating software maintenance cost. The software maintenance cost is expressed by Sheldon [36] and Dhillon [37] as

$$SMC = 3(C_{mm})(n)/\theta, \tag{9.1}$$

where
   SMC is the software maintenance cost.
   C_{mm} is the cost per human-month.
   n is the number of instructions to be changed per month.
   θ is the difficulty constant; its specified values for hard, medium, and easy
      programs are 100, 250, and 500, respectively.

9.6 RCM Goals and Principles

There are many goals of RCM. The important ones are as follows [38]:

- To develop PM-associated tasks that can effectively reinstate safety and reliability to their inherent levels in the event of system/equipment deterioration
- To develop design-related priorities that can effectively facilitate PM
- To collect information considered useful for improving the design of items with proven unsatisfactory inherent reliability
- To achieve all the previously mentioned goals when the total cost is minimal

There are many principles of RCM. Some of them are as follows [8]:

- *RCM is function oriented.* It means that RCM plays an instrumental role in preserving system/equipment function, not just operability for its own sake.
- *RCM tasks must be effective.* It means that the tasks must be technically sound and cost effective.
- *RCM is equipment/system focused.* It means that RCM is more concerned with maintaining system function as opposed to maintaining function of individual component.
- *An unsatisfactory condition is defined as a failure by RCM.* It means that a failure could be either a loss of acceptable quality or a loss of function.

- *Safety and economics drive RCM.* It means that safety is of paramount importance; thus, it must be ensured properly at any cost, and then, cost effectiveness becomes the criterion.

- *Three types of maintenance tasks along with run-to-failure tasks are acknowledged by RCM.* These tasks are known as failure finding, time directed, and condition directed. The purpose of failure-finding tasks is to find hidden functions that have failed without providing any indication of pending failure. Time-directed tasks are scheduled as considered appropriate. Condition-directed tasks are carried out as the conditions indicate for their necessity. Run-to-failure is a conscious decision in RCM.

- *RCM uses a logic tree for screening maintenance tasks.* This provides consistency in the maintenance of all types of equipment.

- *RCM is reliability centered.* It means that RCM is not overly concerned with simple failure rate, but it places importance on the relationship between failures experienced and operating age. In short, RCM treats failure statistics in an actuarial fashion.

- *RCM tasks must be applicable.* It means that tasks must reduce failures or ameliorate secondary damage resulting from failure.

- *Design-related limitations are acknowledged by RCM.* It means that the goal of RCM is to maintain the inherent reliability of the system/ equipment design and, at the same time, recognize that any changes in inherent reliability can only be made through design rather than maintenance. More clearly, maintenance at the best of times can only achieve and maintain a level of designed reliability.

9.7 RCM Process

The RCM process is applied for determining specific maintenance tasks to be carried out as well as for influencing item reliability and maintainability during design. Initially, the RCM process is applied during the design and development phase and then reapplied, as necessary, during the operational phase for sustaining an effective maintenance program based on experience in the field environment.

The basic RCM process is composed of the following seven steps [39]:

- *Step 1: Identify important items in regard to maintenance*—Generally, maintenance-important items are highlighted with the aid of methods such as FTA and failure, mode, effects, and criticality analysis.

- *Step 2: Obtain essential failure data*—In determining occurrence probabilities and assessing criticality, the availability of data on operator error probability, inspection efficiency, and part failure rate is very important. These types of data come from generic failure databanks, field experience, etc.

- *Step 3: Develop FTA data*—The occurrence probabilities of fault events—basic, intermediate, and top events—are estimated as per combinatorial properties of the logic elements in the fault tree.

- *Step 4: Apply decision logic to critical failure modes*—The decision logic is designed to lead, by asking standard assessment-related questions, to the most desirable task combinations of PM. The same logic is applied to each and every crucial failure mode of each maintenance-important item.

- *Step 5: Categorize maintenance requirements*—Maintenance-related requirements are classified under three categories: hard-time maintenance requirements, on-condition maintenance requirements, and condition monitoring maintenance requirements.

- *Step 6: Implement RCM decisions*—Task intervals and frequencies are set/enacted as part of the overall maintenance strategy/plan.

- *Step 7: Apply sustaining engineering on the basis of real-life field experience*— Once the equipment/system starts operating, the real-life field data start to accumulate. At that time, one of the most important steps is to reevaluate all RCM-related default decisions.

9.8 Elements of RCM

There are four major elements of RCM. These elements are reactive maintenance, PM, predictive testing and inspection (PTI), and proactive maintenance [4,10,40]. Each of these elements is described in the following, separately.

9.8.1 Reactive Maintenance

Reactive maintenance is also referred to as repair, *run-to-failure, breakdown,* or *fix-when-fail maintenance*. In this type of maintenance, it is assumed that there is an equally likely chance for a failure occurrence in any component, part, or system. When using this maintenance approach, equipment/system maintenance, repair, or replacement takes place only when deterioration in the condition of an item/equipment/system leads to a functional failure.

When reactive maintenance is practiced solely, poor use of maintenance-related effort, a high replacement of part inventories, and a high percentage of unplanned activities are generally typical. In addition, a totally reactive maintenance program overlooks opportunities to influence equipment/system/item survivability.

It is to be noted that reactive maintenance can be practiced effectively only if it is carried out as a conscious decision, based on the conclusions of an RCM analysis that compares risk and cost of failure with the cost of maintenance needed for mitigating that risk and failure cost. A criterion to determine the priority of repairing or replacing the failed equipment/item in the reactive maintenance program is available in the studies by Dhillon [4] and National Aeronautics and Space Administration [10].

9.8.2 Preventive Maintenance

PM is also known as interval-based or *time-driven maintenance* and is carried out with regard to equipment condition. It consists of periodically scheduled inspection, adjustments, lubrication, part replacement, cleaning, calibration, and repair of components/items. PM schedules regular maintenance and inspection at set intervals for reducing failures for susceptible equipment/system.

It is to be noted that, depending on the predefined intervals, practicing PM can result in a significant increase in routine maintenance and inspections. However, on the other hand, it can help reduce the severity and the frequency of unplanned failures. If PM is the only type of maintenance practiced, it can be quite costly and ineffective. Additional information on PM is available in the study by Dhillon [4].

9.8.3 Predictive Testing and Inspection

PTIs are also known as *predictive maintenance* or *condition monitoring*. In order to assess equipment/item condition, it uses performance data, visual inspection, and nonintrusive testing methods. PTI replaces arbitrarily timed maintenance tasks with maintenance and is carried out as warranted by the equipment/item condition. Analysis of equipment/item condition monitoring data on a continuous basis is very useful to plan and schedule repair/maintenance in advance of functional/catastrophic failure.

The collected PTI data are used for determining the equipment condition and for identifying the failure precursors in many ways, including pattern recognition, statistical process analysis, trend analysis, tests against limits and ranges, data comparison, and correlation of multiple technologies. Finally, it is to be noted that PTI should not be the only type of maintenance practiced, because it does not lend itself to all types of equipment/item or possible failure modes.

9.8.4 Proactive Maintenance

Proactive maintenance is useful for improving maintenance through actions such as better workmanship, design, maintenance procedures, installation, and scheduling. The characteristics of proactive maintenance include items such as follows [4]:

- Optimizing and tailoring maintenance methods and technologies to each application
- Practicing a continuous process of improvement
- Using feedback and communications to ensure that changes in design/procedures are efficiently made available to item designers/ management
- Ensuring that nothing that affects maintenance occurs in total isolation, with the ultimate goal of correcting the concerned equipment forever

Proactive maintenance integrates functions with support maintenance into maintenance program planning, performs root cause failure analysis and predictive analysis for enhancing maintenance effectiveness, uses a life cycle view of maintenance and supporting functions, and conducts periodic evaluation of the technical content and the performance interval of maintenance tasks [10]. The following eight basic methods are used by proactive maintenance to extend item/equipment life [4,10]:

- Root cause failure analysis
- Failed item analysis
- Age exploration (AE)
- Reliability engineering
- Precision rebuild and installation
- Rebuild certification/verification
- Recurrence control
- Specifications for new/rebuilt item/equipment

The first four of the previously mentioned methods are described in the following sections, and the information on the remaining four methods is available in the study by Dhillon [4].

9.8.4.1 Root Cause Failure Analysis

Root cause failure analysis is concerned with proactively seeking the fundamental causes of equipment/facility failure. The main objectives of performing root cause failure analysis are as follows:

- To determine the cause of a problem economically and efficiently
- To rectify the problem cause, not just its effect
- To provide data that can be useful to eradicate the problem
- To instill a mentality of *fix forever*

9.8.4.2 Failed Item Analysis

Failed item analysis involves visually inspecting failed items after removal to determine failure causes. As the need arises, more detailed technical analysis is conducted for finding the real cause of a failure. For example, in the case of bearings, the root causes for the occurrence of their failures may relate to factors such as poor installation, improper lubrication practices, poor storage and handling methods, or excessive balance and alignment tolerances.

Experiences, over the years, indicate that over 50% of all bearing-related problems are due to improper installation or contamination. Generally, indicators of improper installation-related problems are evident on bearings' external and internal surfaces and the indicators of contamination appear on the internal surfaces of bearings.

9.8.4.3 Age Exploration

AE is an important factor in developing an RCM program. AE provides a mechanism for varying key aspects of a maintenance program to optimize the process. The AE method examines the applicability of all maintenance tasks in regard to the three factors presented in the following.

- *Factor 1: Performance interval*—Adjustments are carried out continually to the interval of task performance until the rate at which resistance to failure declines is effectively determined or sensed.
- *Factor 2: Technical content*—The technical contents of a task are examined for ensuring that all highlighted modes of failures are appropriately addressed, as well as assuring that the current maintenance tasks result in the expected degree of reliability.
- *Factor 3: Task grouping*—Tasks with similar periodicity are categorized or grouped for the purpose of improving the total time spent on the job site as well as lowering outages.

9.8.4.4 Reliability Engineering

Reliability engineering, in conjunction with other proactive maintenance methods, involves the modification, the improvement, or the redesign of parts/items or their replacement with better parts/items. In certain cases, a total redesign of the part/item may be required. There are many methods used in reliability engineering for conducting reliability analysis of engineering systems/items. The two most widely used methods in the industry are FMEA and FTA.

Additional information on FMEA and FTA is available in Chapter 4.

9.9 RCM Program Effectiveness Measurement Indicators

In order to measure the effectiveness of an RCM program, over the years, many management indicators have been developed. The numerical indicators/metrics are considered very helpful because they are quantitative, objective, more easily trended than words and precise, in addition to consisting of a benchmark and a descriptor. A benchmark is a numerical expression of a set goal, and a descriptor may be expressed as a word or a group of words detailing the function, the units, and the process under consideration for measurement.

Six indicators/metrics considered useful to measure the effectiveness of a RCM program are presented in the following sections along with their suggested benchmark values [4,10]. These benchmark values are the mean values of data surveyed from approximately 50 major corporations in the early 1990s [10].

9.9.1 Indictor I: Emergency Percentage Index

Emergency percentage index is expressed by

$$P_e = \frac{HWEJ}{HW},$$

(9.2)

where
P_e is the emergency percentage.
HW is the total number of hours worked.
$HWEJ$ is the total number of hours worked on emergency jobs.

For this index, the benchmark figure is 10% or less.

9.9.2 Indicator II: Maintenance Overtime Percentage Index

Maintenance overtime percentage index is defined by

$$P_0 = \frac{NMOH}{NRMH},$$

(9.3)

where
P_0 is the maintenance overtime percentage.
$NRMH$ is the total number of regular maintenance hours during period.
$NMOH$ is the total number of maintenance overtime hours during period.

For this index, the benchmark figure is 5% or less.

9.9.3 Indicator III: Equipment Availability

Equipment availability is expressed by

$$AV_e = \frac{HEEAV}{HDRP},$$ (9.4)

where

AV_e is the equipment availability.
$HDRP$ is the total number of hours during the reporting period.
$HEEAV$ is the number of hours that each unit of equipment is available to run at capacity.

For this metric, the benchmark figure is 96%.

9.9.4 Indicator IV: PM/PTI–Reactive Maintenance Index

PM/PTI–reactive maintenance index is an indicator that is divided into two areas: PM/PTI and reactive maintenance. The PM/PTI-related index is defined by

$$P_{pp} = \frac{MHPPW}{MHPPW + MHRMW},$$ (9.5)

where

P_{pp} is the PM/PTI work percentage.
$MHRMW$ is the total human-hours of reactive maintenance work.
$MHPPW$ is the total human-hours of PM/PTI work.

For this index, the benchmark figure is 70%.
The reactive maintenance-related index is expressed by

$$P_{rm} = \frac{MHRMW}{MHPPW + MHRMW},$$ (9.6)

where

P_{rm} is the reactive maintenance work percentage.

For this index, the benchmark figure is 30%. It is to be noted that the sum of Equations 9.5 and 9.6 is equal to unity or 100%.

9.9.5 Indicator V: Emergency-PM/PTI Work Index

Emergency-PM/PTI work index is defined by

$$P_{epp} = \frac{EWH}{PPWH},$$ (9.7)

where
P_{epp} is the percent of emergency work to PTI and PM work.
$PPWH$ is the total number of PTI and PM work hours.
EWH is the total number of emergency work hours.

For this metric, the benchmark figure is 20% or less.

9.9.6 Indicator VI: PTI-Covered Equipment Index

PTI-covered equipment index is a metric that is used for calculating the percentage of candidate equipment covered by PTI and is defined by

$$P_{ce} = \frac{EIPP}{NECP},$$ (9.8)

where
P_{ce} is the percent of candidate equipment covered by PTI.
$NECP$ is the total number of equipment candidates for PTI.
$EIPP$ is the total number of equipment items in PTI program.

For this index, the benchmark figure is 100%.

9.10 Reasons for RCM Failures and Benefits of RCM

Occasionally, the application of RCM has resulted in failure due to various reasons. Some of these reasons are as follows [6]:

- The application was superfluous or hurried.
- Only one person was assigned to apply RCM.
- An analysis was performed at too low a level.
- Computers were used to drive the process.
- Manufacturers/equipment vendors were asked to apply RCM on their own.

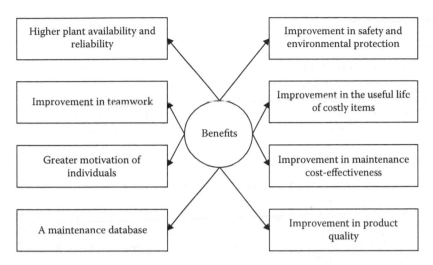

FIGURE 9.2
RCM benefits.

- Too much emphasis placed on failure data.
- Only the maintenance department on its own applied RCM.

There are many benefits of RCM. Some of them are shown in Figure 9.2 [6,39].

PROBLEMS

1. List at least seven facts and figures, directly or indirectly, concerned with software maintenance.
2. Discuss software maintenance problems.
3. What are the types of software maintenance?
4. Describe the following two software maintenance methods:
   a. Impact analysis
   b. Software configuration management
5. What are the factors that affect software maintenance costs?
6. What are the important goals of RCM?
7. Describe the RCM process.
8. What are the elements of RCM? Describe each element in detail.
9. What are the advantages of RCM?
10. List at least eight basic methods used by proactive maintenance to extend item/equipment life.

References

1. Omdahl, T.P., Editor, *Reliability, Availability, and Maintainability (RAM) Dictionary*, ASQC Press, Milwaukee, WI, 1988.
2. IEEE-std-610.12-1990, *IEEE Standard Glossary of Software Engineering Terminology*, Institute of Electrical and Electronic Engineers, New York, 1991.
3. Martin, J., *Fourth-Generation Languages*, Vol. 1, Prentice-Hall, Englewood Cliffs, NJ, 1985.
4. Dhillon, B.S., *Engineering Maintenance: A Modern Approach*, CRC Press, Boca Raton, FL, 2002.
5. Picknell, J., Steel, K.A., Using a CMMS to support RCM, *Maintenance Technology*, October 1997, pp. 110–117.
6. Moubray, J., *Reliability Centered Maintenance*, Industrial Press, New York, 1997.
7. August, J., *Applied Reliability Centered Maintenance*, PennWell, Tulsa, OK, 1999.
8. Nowlan, F.S., Heap, H.F., *Reliability Centered Maintenance*, Dolby Access Press, San Francisco, 1978.
9. Smith, A.M., *Reliability Centered Maintenance*, McGraw-Hill, New York, 1993.
10. *Reliability Centered Maintenance Guide for Facilities and Collateral Equipment*, National Aeronautics and Space Administration, Washington, DC, 1996.
11. Dhillon, B.S., *Engineering Maintainability*, Gulf Publishing Company, Houston, TX, 1999.
12. Horowitz, B.M., *Strategic Buying for the Future*, Libbey Publishing, Washington, DC, 1993.
13. Foster, J., *Cost Factors in Software Maintenance*, PhD dissertation, Department of Computer Science, University of Durham, Durham, 1993.
14. Boeing Company, Software cost measuring and reporting, ASD Document No. D180-22813-1, US Air Force, Washington, DC, January 1979.
15. Schatz, W., Fed facts, *Datamation*, Vol. 15, August 1986, pp. 72–73.
16. Fairley, R.E., *Software Engineering Concepts*, McGraw-Hill, New York, 1985.
17. Coggins, J.M., *Team Software Engineering and Project Management*, Department of Computer Science, University of North Carolina, Chapel Hill, NC, 1994.
18. Stevenson, C., *Software Engineering Productivity*, Chapman and Hall, London, 1995.
19. Glass, R.L., *Software Reliability Guidebook*, Prentice-Hall, Englewood Cliffs, NJ, 1981.
20. Charette, R.N., *Software Engineering Environments*, Intertext Publications, New York, 1986.
21. Parikh, G., Zvegintzov, N., *Tutorial on Software Maintenance*, IEEE Computer Society Press, Los Alamitos, CA, 1993.
22. Lientz, B.P., Swanson, E.B., Problems in application software maintenance, *Communication of the ACM*, Vol. 24, No. 11, 1981, pp. 763–769.
23. Bennett, K.H., *Software Maintenance: A Tutorial, in Software Engineering*, edited by M. Dorfman, R.H., Thayer, IEEE Computer Society Press, Alamitos, CA, 1997, pp. 289–303.
24. Pfleeger, S.L., *Software Engineering: Theory and Practice*, Prentice-Hall, Upper Saddle River, NJ, 1998.
25. *Pricing Handbook*, Federal Aviation Administration, Washington, DC, 1999.

26. Pfleeger, S.L., Bohner, S., A framework for maintenance metrics, *Proceedings of the IEEE Conference on Software Maintenance*, 1990, pp. 225–230.
27. Schneider, G.R.E., Structured software maintenance, *Proceedings of the AFIPS National Computer Conference*, 1983, pp. 137–144.
28. Yourdon, E., Structured maintenance, in *Techniques of Program and System Maintenance*, edited by G. Parikh, Ethnotech, Lincoln, NE, 1980, pp. 211–213.
29. Hall, R.P., Seven ways to cut software maintenance costs, *Datamation*, July 15, 1987, pp. 81–84.
30. Arthur, L.J., *Software Evolution: The Software Maintenance Challenge*, Wiley, New York, 1988.
31. Lindhorst, W.M., Scheduled maintenance of applications software, *Datamation*, May 1973, pp. 64–67.
32. Cashman, P.M., Holt, A.W., A communication-oriented approach to structuring the software maintenance environment, *ACM SIGSOFT Software Engineering Notes*, Vol. 5, No. 1, January 1, 1980, pp. 14–17.
33. Shooman, M.L., *Software Engineering*, McGraw-Hill, New York, 1983.
34. Mills, H.D., Software development, *Proceedings of the IEEE Second International Conference on Software Engineering*, Vol. 11, 1976, pp. 79–83.
35. Belady, L., Lehman, M.M., An introduction to growth dynamics, in *Statistical Computer Performance Evaluation*, edited by W. Freiberger, Academic Press, New York, 1972.
36. Sheldon, M.R., *Life Cycle Costing: A Better Method for Government Procurement*, Westview Press, Boulder, CO, 1979.
37. Dhillon, B.S., *Life Cycle Costing*, Gordon and Breach Science Publishers, New York, 1989.
38. AMC Pamphlet No. 750-2, *Guide to Reliability Centered Maintenance*, Department of the Army, Washington, DC, 1985.
39. Brauer, D.C., Brauer, G.D., Reliability-centered maintenance, *IEEE Transactions on Reliability*, Vol. 36, 1987, pp. 17–24.
40. Report No. NAVAIR 00-25-403, *Guidelines for the Naval Aviation Reliability-Centered Maintenance Process*, Naval Air Systems Command, Department of Defense, Washington, DC, October 1996.

10

Maintenance Safety and Human Error in Aviation and Power Plant Maintenance

10.1 Introduction

Each year, billions of dollars are being spent on maintenance for keeping engineering systems and items in operational state; the problem of safety in maintenance has become a very important issue. For example, in 1994, in the US mining industrial sector, 13.61% of all accidents occurred during the maintenance process, and since 1990, the occurrence of such accidents has been increasing annually [1]. The problem of safety in maintenance activity is not only for ensuring the safety of the maintenance workforce, but also for ensuring the safety-related actions taken by these people. For example, an incorrect action taken by aircraft maintenance personnel can lead to a loss of many lives.

Maintenance is an important element of the aviation industrial sector throughout the world, and in 1989, US airlines spent approximately 12% of their operating costs on the maintenance-related activities [2,3]. During the period from 1980 to 1988, the airline maintenance cost increased from around US$2.9 billion to US$5.7 billion due to factors such as increase in air traffic and increased maintenance for continuing aircraft worthiness of aging aircraft [4].

Over the years, increase in air traffic and increased demands on aircraft utilization due to the stringent requirements of commercial schedules continue to put pressures on the maintenance-related activities for on-time performance. In turn, this has increased chances for the human error occurrence in aircraft maintenance-related operations [5]. A study performed in the United Kingdom reported that during the period from 1990 to 2000, the occurrence of maintenance error-related events per million flights has doubled [6].

In power plants, maintenance is an important activity, and it consumes a significant sum of money spent on power generation. In the causation of power generation safety-related incidents, human error in maintenance has been found to be an important factor [7]. For example, a study of reliability-related events concerning electrical/electronic components in nuclear power plants (NPPs) reported that human error made by maintenance workers exceeded operator errors and more than three quarters of the errors took place during the testing and the maintenance activity [7,8].

This chapter presents various important aspects of maintenance safety and human error in aviation and power plant maintenance.

10.2 Maintenance Safety-Related Facts, Figures, and Examples

Some of the facts, the figures, and the examples directly or indirectly concerned with maintenance safety are as follows [9,10]:

- As per the study by National Safety Council [1], around 3.8 million workers in the United States suffered from disabling injuries on the job in 1998.
- Each year, the US industry spends around US$300 billion on plant maintenance and operations [11].
- A study of safety issues in regard to onboard fatality of worldwide jet fleet reported that, for the period of 1982–1991, maintenance and inspection were the second most important safety-related issues with a total of 1481 onboard fatalities [12,13].
- In 1994, 13.6% of the accidents in the US mining industrial sector took place during maintenance.
- In 1985, 520 fatalities occurred in a Japan Airlines Boeing 747 jet accident due to an improper repair [14,15].
- As per the study by Goetsch [16], in 1991, an explosion caused four fatalities in an oil refining company in Louisiana, and it occurred as three gasoline-synthesizing units were being put into operation after some maintenance activities.
- As per the study by Christensen and Howard [17], an incident at the Ekofish Oil Field in the North Sea involving the blowout preventer (assembly of valves) was due to upside-down installation of the device, and its estimated cost was approximately $50 million.
- In 1990, due to a steam leak in the fire room, 10 fatalities occurred on the USS *Iwo Jima* (LPH2) naval ship. A subsequent investigation into the accident revealed that maintenance workers just repaired a valve and replaced bonnet fasteners with mismatched and incorrect material [18].
- As per the studies by the Australian Transport Safety Bureau [14] and the Ministry of Transportation [19], in 1990, a newly replaced

windscreen of a British BAC 1-11 jet blew out as the aircraft was climbing to its cruising altitude due to the wrong installation of the windscreen by a maintenance worker.

10.3 Factors Responsible for Dubious Safety Reputation in Performing Maintenance Tasks and Reasons for Safety-Related Problems in Maintenance

There are many factors responsible for the dubious safety reputation in performing maintenance tasks. Some of these factors are as follows [20]:

- Sudden need for maintenance work, thus allowing a limited time for necessary preparation
- Performance of maintenance tasks underneath/inside items such as air ducts, pressure vessels, and large rotating machines
- Performance of maintenance tasks at rather odd hours, in small numbers, and in remote locations
- Disassembling previously functioning items, thus working under the risk of releasing stored energy
- Difficulty in keeping regular communication with personnel involved in maintenance tasks
- Maintenance tasks carried out in unfamiliar surroundings/territories implying that hazards such as rusted handrails, missing gratings, and broken light fittings may go unnoticed
- Need for carrying heavy and bulky items from a warehouse/store to the maintenance site, sometimes using transport and lifting equipment way beyond the boundaries of a strict maintenance regime
- Frequent occurrence of numerous maintenance-related tasks (e.g., machinery failures), thus fewer opportunities for discerning safety-related problems as well as for introducing remedial measures
- Time-to-time maintenance tasks requiring carrying out of tasks such as rough handling of rather cumbersome heavy items in confined spaces and poorly lit areas or disassembling corroded parts.

Experiences, over the years, clearly indicate that there is a significant proportion of accidents that occur during maintenance. Some of the important reasons for safety-associated problems in maintenance are as shown in Figure 10.1 [9,10].

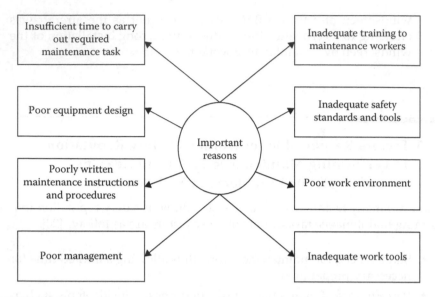

FIGURE 10.1
Important reasons for safety-associated problems in maintenance.

10.4 Maintenance Personnel Safety and Maintenance Safety-Related Questions for Manufacturers of Engineering Systems/Equipment

Generally, during the design phase of an engineering equipment/system, emphasis is placed on designing the safety of the equipment/system rather than on the safety of personnel such as maintainers and operators. Experiences, over the years, clearly indicate that more time-to-time protection is required for maintenance workers beyond the safety-designed equipment/systems. In this regard, two important areas concerning the safety of maintenance workers are respiratory protection and protective clothing [20].

Some of the important areas for respiratory protection are shown in Figure 10.2.

Some of the important items of protective clothing are as follows:

- *Helmets and hard hats:* Helmets and hard hats are essential for protecting maintenance personnel from head injury.
- *Ear defenders:* Ear defenders are essential for protecting maintenance personnel from damaging their ears in an environment where excessive noise is generated by processes or machines.

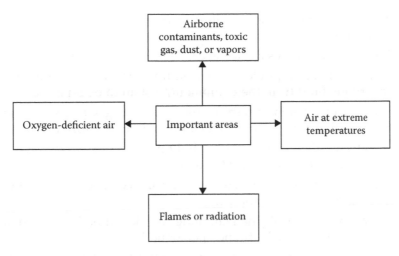

FIGURE 10.2
Important areas for respiratory protection.

- *Boots and toecaps:* Boots and toecaps are essential for reducing injury risk in situations such as dismantling used equipment where heavy metal items are difficult to hold and are quite likely to slip and drop on involved maintenance personnel's exposed feet.

There are many areas in which manufacturers of equipment/systems can, directly or indirectly, play an important role in improving maintenance safety during equipment/system field operation. Questions such as presented in the following can be useful to manufacturers for determining whether the common problems that might be encountered during the maintenance phase have been addressed properly [21]:

- Are the units/parts requiring frequent maintenance easily accessible?
- Do the instructions incorporate proper warnings for alerting maintenance personnel of any danger?
- Is the repair process hazardous to all concerned repair workers?
- Can the disassembled equipment/system for repair be reassembled incorrectly so that it becomes hazardous to all potential users?
- Were human factor principles properly applied for minimizing maintenance-related problems?
- Do the repair instructions contain proper warnings for wearing appropriate protective gear because of pending hazards?
- Does the equipment contain proper safety interlocks that must be bypassed for performing required repairs/adjustments?

- Are satisfactorily and well-written instructions available for maintenance and repair?
- Are all the test points located at easy-to-reach locations?
- Is there an appropriate system/equipment for removing fuel/hazardous fluid from the equipment/system to be repaired?
- Was proper attention given for reducing voltages to levels at test points so that hazards to maintenance workers are minimized?
- Does the equipment contain any built-in mechanism for indicating that safety-critical items need maintenance?
- Are the proper warnings against working placed on items that can shock maintenance personnel?
- Is it possible to repair the item under consideration by individuals other than the specially trained personnel?
- Is there an appropriate mechanism installed for indicating when the redundant units of safety-critical systems fail?
- Is the need for special tools for repairing safety-critical items minimized to an acceptable level?
- Is the equipment/system designed in such a way so that after experiencing a failure, it would automatically stop functioning and will not cause any damage?

10.5 Guidelines for Equipment/System Designers for Improving Safety in Maintenance

Over the years, professionals working in the area of maintenance have developed various useful guidelines for equipment/system designers for improving safety in maintenance. Some of these guidelines are presented in the following [21]:

- Eliminate or reduce the opportunity for performing adjustments or maintenance close to hazardous functioning items.
- Develop designs/procedures in such a way that the maintenance error occurrence probability is minimized.
- Eliminate or reduce the need for special tools.
- Simplify the design as much as possible because complex designs generally add to maintenance-related problems.
- Provide effective guards against moving items.

- Develop the design in such a way that the chances of maintenance workers being injured by escaping high-pressure gas, electric shock, etc., are minimized.
- Design for appropriate accessibility so that items requiring maintenance are easy and safe to service, remove, replace, or check.
- Incorporate appropriate interlocks for blocking access to hazardous locations.
- Incorporate proper fail-safe designs for preventing damage or injury when a failure occurs.
- Incorporate appropriate devices or other appropriate measures for allowing early detection or prediction of potential failures/faults so that proper maintenance can be carried out prior to actual failure with a reduced risk of hazard.
- Pay careful attention to typical behaviors of humans.

10.6 Models for Performing Maintenance Safety Analysis

Over the years, numerous mathematical models have been developed for performing various types of reliability analysis [22]. Some of these models can equally be used for performing maintenance safety analysis. Two such models are presented in the following section [10].

10.6.1 Model I

Model I is a mathematical model that represents an engineering system having three states: operating normally, operating unsafely, and being failed. The system is repaired from unsafe operating state and failed state. The system transition diagram is shown in Figure 10.3. The numerals in the circle, the rectangle, and the diamond denote system states. The Markov method is used to develop equations for system state probabilities and MTTF.

The following assumptions are associated with this model:

- Failures occur independently.
- The repaired system is as good as new.
- System failure and repair rates are constant.

The following symbols are associated with this model:

j is the jth state of the system; $j = 0$ means that the system is operating normally, $j = 1$ means that the system is operating unsafely, and $j = 2$ means that the system failed.

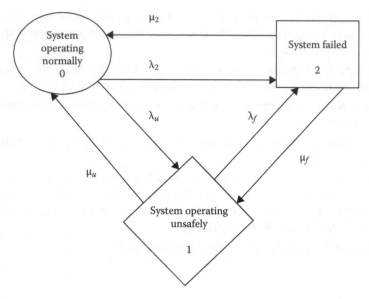

FIGURE 10.3
System transition diagram.

$P_j(t)$ is the probability that the system is in state j at time t; for $j = 0, 1, 2$.
t is the time.

λ_j is the system jth failure rate; $j = u$ means from state 0 to state 1; $j = 2$ means from state 0 to state 2; and $j = f$ means from state 1 to state 2.

μ_j is the system jth repair rate; $j = u$ means from state 1 to state 0, $j = 2$ means from state 2 to state 0, and $j = f$ means from state 2 to state 1.

With the aid of the Markov method, we get the following differential equations for the Figure 10.3 diagram [10,22]:

$$\frac{dP_0(t)}{dt} + (\lambda_2 + \lambda_u)P_0(t) = \mu_u P_1(t) + \mu_2 P_2(t), \tag{10.1}$$

$$\frac{dP_1(t)}{dt} + (\mu_u + \lambda_f)P_1(t) = \mu_f P_2(t) + P_0(t)\lambda_u, \tag{10.2}$$

$$\frac{dP_2(t)}{dt} + (\mu_2 + \mu_f)P_2(t) = \lambda_f P_1(t) + \lambda_2 P_0(t). \tag{10.3}$$

At time $t = 0$, $P_0(0) = 1$ and $P_1(0) = P_2(0) = 0$.

For a very large t, by solving Equations 10.1 through 10.3, we get the following steady-state probabilities [22]:

$$P_0 = \frac{(\mu_2 + \mu_f)(\mu_u + \lambda_f) - \lambda_f \mu_f}{X}, \tag{10.4}$$

where

$$X = (\mu_2 + \mu_f)(\mu_u + \lambda_u + \lambda_f) + \lambda_2(\mu_u + \lambda_f) + \lambda_2\mu_f + \lambda_u\lambda_f - \lambda_f\mu_f$$

$$P_1 = \frac{\lambda_u(\mu_2 + \mu_f) + \lambda_2\mu_f}{X}, \tag{10.5}$$

$$P_2 = \frac{\lambda_2\lambda_f + \lambda_2(\mu_u + \lambda_f)}{X}, \tag{10.6}$$

where
P_0, P_1, and P_2 are the steady-state probabilities of the system being in states 0, 1, and 2, respectively.

The steady-state probability of the system operating unsafely is given by Equation 10.5.

By setting $\mu_2 = \mu_f = 0$ in Equations 10.1 through 10.3 and solving the resulting equations, we get the following expression for the system reliability:

$$R_s(t) = P_0(t) + P_1(t)$$
$$= (X_1 + Y_1)e^{r_1 t} + (X_2 + Y_2)e^{r_2 t}, \tag{10.7}$$

where
$R_s(t)$ is the system reliability at time t.

$$r_1 = \frac{-N_1 + \sqrt{N_1^2 - 4N_2}}{2}.$$

$$r_2 = \frac{-N_1 - \sqrt{N_1^2 - 4N_2}}{2}.$$

$$N_1 = \mu_u + \lambda_2 + \lambda_u + \lambda_f.$$
$$N_2 = \lambda_2\mu_u + \lambda_2\lambda_f + \lambda_u\lambda_f.$$
$$X_1 = \frac{r_1 + \mu_u + \lambda_f}{(r_1 - r_2)}.$$

$$X_2 = \frac{r_2 + \mu_u + \lambda_f}{(r_2 - r_1)}.$$

$$Y_1 = \frac{\lambda_u}{(r_1 - r_2)}.$$

$$Y_2 = \frac{\lambda_u}{(r_2 - r_1)}.$$

By integrating Equation 10.7 over the time interval $[0, \infty]$, we obtain the following equation for the system MTTF [10,22].

$$MTTF_s = \int_0^\infty R_s(t)\,dt$$

$$= -\left[\frac{(X_1 + Y_1)}{r_1} + \frac{(X_2 + Y_2)}{r_2} \right]. \tag{10.8}$$

Example 10.1

Assume that a system can be either operating normally, operating unsafely, or failed. Its failure rates from normal operating state to unsafe operating state, unsafe operating state to failed state, and normal operating state to failed state are 0.003, 0.001, and 0.02 failures/hour, respectively. Furthermore, the system repair rates from the failed state to the normal operating state, the unsafe operating state to the normal operating state, and the failed state to the unsafe operating state are 0.05, 0.005, and 0.008 repairs/hour, respectively.

Calculate the probability of the system being in unsafe operating state during a very large mission period by using the specified data values.

By substituting the given data values into Equation 10.5, we get

$$P_1 = \frac{(0.003)(0.05 + 0.008) + (0.02)(0.008)}{X}$$

$$= 0.4191,$$

where

$$X = (0.05 + 0.008)(0.005 + 0.003 + 0.001) + (0.02)(0.005 + 0.001)$$
$$+ (0.02 + 0.008) + (0.003 + 0.001) - (0.001)(0.008)$$

Thus, there is approximately 42% chance that the system will be operating in unsafe state during a very large mission period.

10.6.2 Model II

Model II is a mathematical model that represents an engineering system having three states: operating normally, failed unsafely, and failed safely. The failed system is repaired. The system transition diagram is shown in Figure 10.4.

The numerals in the circle, the rectangle, and the diamond denote system states. The Markov method is used to develop equations for system state probabilities (i.e., operating normally, failed safely, and failed unsafely) and MTTF.

The following assumptions are associated with the Figure 10.4 diagram model:

- Failures occur independently.
- The system can fail either unsafely or safely.
- The system failure and repair rates are constant.
- The repaired system is as good as new.

The following symbols are associated with this model:

j is the jth state of the system; $j = 0$ means that the system is operating normally; $j = 1$ means that the system failed safely; and $j = 2$ means that the system failed unsafely.

$P_j(t)$ is the probability that the system is in state j at time t; for $j = 0, 1, 2$.

t is the time.

λ_j is the system jth failure rate; $j = 1$ means safe, and $j = 2$ means unsafe.

μ_j is the failed system jth repair rate; $j = 1$ means from safe failed state, and $i = 2$ means from unsafe failed state.

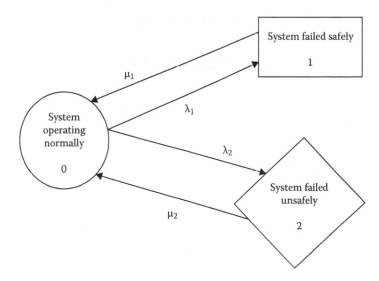

FIGURE 10.4
System transition diagram.

With the aid of the Markov method, we get the following differential equations for the Figure 10.4 diagram [10,22]:

$$\frac{dP_0(t)}{dt} + (\lambda_1 + \lambda_2)P_0(t) = \mu_1 P_1(t) + \mu_2 P_2(t), \tag{10.9}$$

$$\frac{dP_1(t)}{dt} + \mu_1 P_1(t) = \lambda_1 P_0(t), \tag{10.10}$$

$$\frac{dP_2(t)}{dt} + \mu_2 P_2(t) = \lambda_2 P_0(t). \tag{10.11}$$

At time $t = 0$, $P_0(0) = 1$ and $P_1(0) = P_2(0) = 0$.

By solving Equations 10.9 through 10.11, we obtain

$$P_0(t) = \frac{\mu_1 \mu_2}{Z_1 Z_2} + \left[\frac{(Z_1 + \mu_1)(Z_1 + \mu_2)}{Z_1(Z_1 - Z_2)} \right] e^{Z_1 t} - \left[\frac{(Z_2 + \mu_1)(Z_2 + \mu_2)}{Z_2(Z_1 - Z_2)} \right] e^{Z_2 t}, \tag{10.12}$$

where

$$Z_1, Z_2 = \frac{-B \pm \sqrt{B^2 - 4(\mu_1 \mu_2 + \lambda_1 \mu_2 + \lambda_2 \mu_2)}}{2}.$$

$$B = \mu_1 + \mu_2 + \lambda_1 + \lambda_2.$$

$$Z_1 Z_2 = \mu_1 \mu_2 + \lambda_1 \mu_2 + \lambda_2 \mu_1.$$

$$Z_1 + Z_2 = -(\mu_1 + \mu_2 + \lambda_1 + \lambda_2).$$

$$P_1(t) = \frac{\lambda_1 \mu_2}{Z_1 Z_2} + \left[\frac{(\lambda_1 Z_1 + \lambda_1 \mu_2)}{Z_1(Z_1 - Z_2)} \right] e^{Z_1 t} - \left[\frac{(\mu_2 + Z_2)\lambda_1}{Z_2(Z_1 - Z_2)} \right] e^{Z_2 t}. \tag{10.13}$$

$$P_2(t) = \frac{\lambda_2 \mu_1}{Z_1 Z_2} + \left[\frac{(\lambda_2 Z_1 + \lambda_2 \mu_1)}{Z_1(Z_1 - Z_2)} \right] e^{Z_1 t} - \left[\frac{(\mu_1 + Z_2)\lambda_2}{Z_2(Z_1 - Z_2)} \right] e^{Z_2 t}. \tag{10.14}$$

Equations 10.13 and 10.14 give the probability of the system failing safely and unsafely, respectively, when subjected to the repair process. As time t

becomes very large, the steady-state probability of the system failing safely using Equation 10.13 is

$$P_1 = \lim_{t \to \infty} P_1(t) = \frac{\lambda_1 \mu_2}{Z_1 Z_2},$$
(10.15)

where
P_1 is the steady-state probability of the system failing safely.

Similarly, as time t becomes very large, the steady-state probability of the system failing unsafely with the aid of Equation 10.14 is

$$P_2 = \lim_{t \to \infty} P_2(t) = \frac{\lambda_2 \mu_1}{Z_1 Z_2},$$
(10.16)

where
P_2 is the steady-state probability of the system failing unsafely.

By setting $\mu_1 = \mu_2 = 0$ in Equations 10.9 through 10.11 and then solving the resulting equations, we obtain

$$P_0(t) = e^{-(\lambda_1 + \lambda_2)t},$$
(10.17)

$$P_1(t) = \frac{\lambda_1}{\lambda_2 + \lambda_1} \left[1 - e^{-(\lambda_1 + \lambda_2)t} \right],$$
(10.18)

$$P_2(t) = \frac{\lambda_2}{\lambda_2 + \lambda_1} \left[1 - e^{-(\lambda_1 + \lambda_2)t} \right].$$
(10.19)

Equation 10.17 is the system reliability/probability of success at time t. In contrast, Equations 10.18 and 10.19 are the probabilities of the system failing safely and unsafely at time t, respectively, without the performance of repair.

By integrating Equation 10.17 over the time interval $[0, \infty]$, we obtain the following equation for system MTTF [22]:

$$MTTF_s = \int_0^\infty P_0(t) dt$$

$$= \int_0^\infty e^{-(\lambda_1 + \lambda_2)t} dt$$
(10.20)

$$= \frac{1}{\lambda_1 + \lambda_2},$$

where

$MTTF_s$ is the system MTTF.

Example 10.2

An engineering system can fail unsafely or safely, and its unsafe and safe failure rates are 0.002 and 0.009 failures/hour, respectively. Furthermore, its unsafe and safe failure mode repair rates are 0.008 and 0.05 repairs/hour, respectively.

Calculate the probability of the engineering system being in unsafe failure mode during a very large mission period.

By inserting the specified data values into Equation 10.16, we get

$$P_2 = \frac{\lambda_2 \mu_1}{Z_1 Z_2} = \frac{(0.002)(0.05)}{(0.05)(0.008) + (0.009)(0.008) + (0.002)(0.05)}$$

$$= 0.1748.$$

Thus, there is an approximately 17% chance that the engineering system will be in unsafe failure mode during a very large mission period.

10.7 Aviation Maintenance Human Error-Related Facts, Figures, and Examples

Some of the aviation maintenance human error-related facts, figures, and examples are as follows:

- Maintenance error contributes to 15% of air carrier accidents and each year costs the US industry over US$1 billion [23].
- A Boeing study reported that 19.2% of in-flight engine shutdowns are due to maintenance error [23].
- A study reported that around 18% of all aircraft accidents are maintenance related [24,25].
- As per the study by Marx and Graeber [26], a study revealed that inspection and maintenance are the factor in about 12% of major aircraft accidents.
- In 1979, 272 deaths occurred in a DC-10 accident due to improper maintenance procedures followed by maintenance workers [27].
- A study of 122 maintenance-related errors that occurred in a major airline over a 3-year period revealed that their breakdowns were

omission (56%), wrong parts (8%), incorrect installations (30%), and other (6%) [28,29].

- An analysis of safety-related issues versus onboard fatalities among jet fleets worldwide during the period of 1982–1991 highlighted inspection and maintenance as the second most important safety-related issue with onboard fatalities [30,31].
- As per the study by the International Civil Aviation Organization [6] and Wenner and Drury [32], in 1988, the upper cabin structure of a Boeing 737-200 aircraft was ripped away during a flight due to structural failure, basically because of failure of maintenance inspectors to highlight over 240 cracks in the aircraft skin during the inspection process.
- In 1991, an Embraer 120 aircraft accident resulted in 13 fatalities due to a human error during scheduled maintenance [5,6].

10.8 Major Categories of Human Errors in Aviation Maintenance and Inspection Tasks and Causes of Human Error in Aviation Maintenance

There are many major/main categories of human errors that occur in aviation maintenance and inspection tasks. Eight of these categories are as follows [28,33,34]:

- Missing part (e.g., bolt–nut not secured)
- Wrong assembly sequence (e.g., incorrect sequence of inner cylinder spacer and lock ring assembly)
- Incorrect part (e.g., wrong pitot-static probes installed)
- Defective part (e.g., cracked pylon, worn cables, fluid leakage)
- Wrong configuration (e.g., valve inserted in backward direction)
- Tactile defects (e.g., seat not locking in right position)
- Procedural defects (e.g., nose landing gear door not closed)
- Functional defects (e.g., incorrect tire pressure)

There are a large number of factors that can, directly or indirectly, impact the performance of personnel involved with aviation maintenance. A document prepared by the International Civil Aviation Organization listed over 300 such factors/influences [35]. These factors/influences range from temperature to boredom.

Some of the important reasons, directly or indirectly, for the occurrence of human error in aviation maintenance are as follows [27,36]:

- Poorly written maintenance procedures
- Fatigued maintenance personnel
- Time pressure
- Outdated maintenance manuals
- Inadequate training, work tools, and experience
- Poor work layout
- Poor work environment (e.g., lighting, temperature, humidity)
- Poor equipment design
- Complex maintenance tasks

10.9 Common Human Errors in Aircraft Maintenance Tasks and Guidelines to Reduce Human Error in Aircraft Maintenance-Related Tasks

A number of studies conducted over the years have identified commonly occurring human errors in aircraft maintenance tasks. One of these studies carried out by the UK Civilian Aviation Authority over a period of 3 years has highlighted the following eight commonly occurring human errors in aircraft maintenance [27,28]:

- Incorrect installation of parts
- Unsecured refuel panels and fuel caps
- Inadequate lubrication
- Fitting of incorrect parts
- Unsecured access panels, cowlings, and fairings
- Failure to remove landing gear ground lock pins prior to aircraft departure
- Discrepancies in electrical wiring including cross connections
- Loose objects/items such as tools left in the aircraft

Over the years, various guidelines have been developed to reduce the occurrence of human error in aircraft maintenance-related tasks. These guidelines cover areas as shown in Figure 10.5 [30,34].

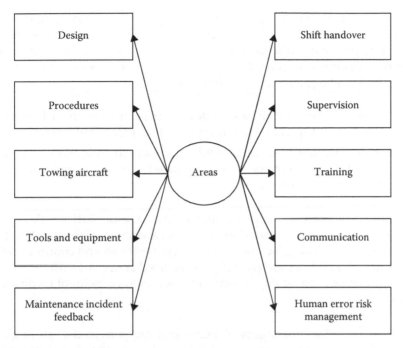

FIGURE 10.5
Areas covered in guidelines to reduce the occurrence of human error in aircraft maintenance tasks.

Two guidelines in the area of design are as follows:

- Ensure that during the design phase, the equipment manufacturers give proper attention to maintenance-related human factors.
- Actively seek relevant information concerning the occurrence of human error during the maintenance phase, to provide useful inputs during the design phase.

A useful guideline in the area of communication is to ensure that appropriate systems are in place for disseminating important information to all personnel concerned with maintenance, so that repeated errors or changing procedures are carefully considered.

Two useful guidelines in the area of training are as follows:

- Consider introducing crew resourcement for individuals concerned with the maintenance activity.
- Provide on a periodic basis training courses to all individuals involved with maintenance with emphasis on company procedures.

A particular guideline in the area of supervision is to recognize that management- and supervision-associated oversights must be strengthened as much as possible, particularly in the final hours of all shifts, as the occurrence of errors becomes more likely. Two useful guidelines pertaining to the area of tools and equipment are as follows:

- Ensure the storage of all lockout devices in such a way that it becomes immediately apparent when they are left in place inadvertently.
- Review carefully systems by which items such as lighting systems and stands are kept for removing unserviceable equipment from service and repairing it quickly.

One important guideline pertaining to the area of shift handover is to ensure that the effectiveness of practices is associated with shift handover by carefully considering factors such as documentation and communication, so that incomplete tasks are correctly transferred across all shifts.

Two guidelines concerning the area of maintenance incident feedback are as follows:

- Ensure that all management personnel are provided with proper feedback on the occurrence of human factor-related maintenance incidents on a regular basis, with consideration to the conditions that play a pivotal role in the occurrence of such incidents.
- Ensure that all personnel associated with the training activity are provided with effective feedback on the occurrence of human factor-related maintenance incidents on a regular basis, so that proper corrective measures aimed at these problems are taken effectively.

Some of the guidelines pertaining to the area of procedures are reviewing all documented maintenance procedures and practices periodically in regard to items such as consistency, realism, and accessibility, ensuring that standard work practices are being followed properly throughout aircraft maintenance operations, and viewing maintenance work practices on a regular basis for ensuring that they do not vary significantly from formal procedures.

An important guideline in the area of towing aircraft is to review the equipment and the procedures used for towing and from maintenance facilities periodically.

Finally, some of the guidelines in the area of human error risk management are to review the need to disturb normally functioning systems to carry out rather nonessential periodic maintenance, because the disturbance may lead to human error; to formally review the effectiveness of defenses such as engine runs built into the system to detect maintenance errors; and to avoid simultaneously carrying out the same maintenance task on similar redundant units.

10.10 Methods for Performing Aircraft Maintenance Error Analysis

Over the years, many methods have been developed in reliability and its related areas that can be used for performing human error analysis in the area of aircraft maintenance. Three of these methods are error–cause removal program (ECRP), cause-and-effect diagram, and fault tree analysis (FTA). The first two methods are described in the following sections, and the application of the fault tree analysis to perform aircraft maintenance error analysis is demonstrated in the study by Dhillon [37].

10.10.1 Error–Cause Removal Program

ECRP was developed to reduce the occurrence of human error to some tolerable level in production operations [38,39]. It can also be used to reduce the occurrence of human error in aircraft maintenance operations. The emphasis of ECRP is on preventive actions rather merely on remedial ones. With respect to aircraft maintenance, this method may simply be described as the maintenance worker participation program to reduce the occurrence of human errors.

More clearly, the ECRP is made up of teams of workers (e.g., aircraft maintenance workers) with each team having its own coordinator, who has special technical and group-associated skills. During meetings, held periodically, workers present their error and error-likely reports. After relevant discussions on these reports, necessary recommendations are made for appropriate remedial or preventive actions. The coordinators of the teams present the recommendations to the management for necessary measures.

The basic elements of this method are as follows [38,39]:

- All personnel involved with this method (i.e., ECRP) are educated about its usefulness.
- All involved maintenance personnel and team coordinators are trained in data collection and analysis methods.
- The aircraft maintenance workers' efforts with respect to ECRP are recognized by the management.
- Human factors and other specialists determine the effects of changes carried out in, say, aircraft maintenance-related operations with the help of the ECRP inputs.
- All proposed solutions are examined in regard to cost by human factors and other specialists.
- The management fully implements the most promising proposed solutions.

- Aircraft maintenance personnel report and evaluate errors and error-likely situations, in addition to proposing solutions for eradicating error causes.

Finally, useful guidelines concerning this method are as follows:

- Restrict to the identification of work conditions that need redesigning to reduce the error occurrence potential.
- Focus on the collection of data on items such as accident-prone conditions, errors, and error-likely conditions.
- Examine each and every work redesign recommended by the team in regard to factors such as the degree of error eradication and increments in cost effectiveness and job satisfaction.

10.10.2 Cause-and-Effect Diagram

The cause-and-effect diagram was developed in the early 1950s by a Japanese man named K. Ishikawa; in the published literature, this diagram is also referred to as the *Ishikawa diagram* or the *fishbone diagram*. This diagram can be a quite useful tool in determining the root causes of a stated aircraft maintenance error and generating appropriate relevant ideas.

Pictorially, the box on the extreme right-hand side of the diagram denotes effect and that on the left-hand side denotes the possible causes that are connected to the centerline. In turn, normally, each cause is made up of various subcauses. Normally, the following steps are followed for developing a cause-and-effect diagram [23]:

- *Step A:* Develop problem statement.
- *Step B:* Brainstorm to highlight possible causes.
- *Step C:* Establish main cause-related categories by stratifying into natural groupings and process steps.
- *Step D:* Develop the diagram by connecting all the possible causes by following the proper process steps and fill in the effect (i.e., the problem) in the box on the extreme right-hand side of the diagram.
- *Step E:* Refine cause categories by raising questions such as why does this condition exist? And what causes this?

Some of the main advantages of the cause-and-effect diagram are as follows:

- A useful method to highlight root causes
- An effective approach to generate ideas
- A useful tool to guide further inquiry
- An effective tool for presenting an orderly arrangement of theories

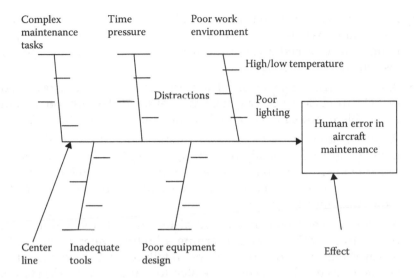

FIGURE 10.6
Cause-and-effect diagram for Example 10.3.

Example 10.3

Assume that there are the following five causes for the occurrence of human error in aircraft maintenance:

- Inadequate tools
- Complex maintenance tasks
- Poor equipment design
- Time pressure
- Poor work environment

The subcauses of the cause *poor work environment* are poor lighting, high/low temperature, and distractions. Draw a cause-and-effect diagram for the effect: human error in aircraft maintenance.

The cause-and-effect diagram for the example is shown in Figure 10.6.

10.11 Power Plant Maintenance Human Error-Related Facts, Figures, and Examples

Some of the facts, the figures, and the examples directly or indirectly concerned with human error in power plant maintenance are as follows:

- As per the study by Hasegawa and Kameda [40], a study of 199 human errors that occurred during the period of 1965–1995 in Japanese NPPs revealed that about 50 of them were concerned with maintenance activities.

- As per the study by Daniels [41], a study revealed that over 20% of all system failures in fossil power plants occur due to human errors and maintenance-related account for around 60% of the annual power loss due to human errors.
- A study of 126 human error-related significant events that occurred in 1990, in nuclear power generation, revealed that around 42% of the problems were linked to maintenance and modification [42].
- As per the study by the Organization for Economic Co-operation and Development [43], a study of NPP operating experiences reported that due to errors in maintenance of some motors in the rod drives, many of the motors ran in a backward direction and withdrew rods, rather than inserting them.
- A study of over 4400 maintenance history records that covered the period from 1992 to 1994, concerning a boiling water reactor NPP, revealed that approximately 7.5% of all failure-related records could be categorized as human errors related to maintenance tasks [44,45].
- A blast at the Ford Rouge power plant in Dearborn, Michigan, that caused six fatalities and injured many workers was due to a maintenance error [46,47].
- In 1989 on Christmas Day, two nuclear reactors were shutdown because of maintenance error and caused rolling blackouts in the state of Florida [48].

10.12 Human Error Causes in Power Plant Maintenance and Most Susceptible Maintenance Tasks to Human Error in Power Generation

There are many causes for the human error occurrence in power plant maintenance. These causes on the basis of characteristics obtained from modeling the maintenance task may be grouped under the following four classifications [7]:

- *Classification 1: Design shortcomings in hardware and software*—Design shortcomings in hardware and software are shortcomings that include items such as wrong or confusing procedures, deficiencies in the design of displays and controls, and insufficient communication equipment.
- *Classification 2: Disturbances of the external environment*—Some examples of disturbances of the external environment are the physical conditions such as temperature, humidity, ventilation, and ambient illumination.

- *Classification 3: Induced circumstances*—Induced circumstances include items such as emergency conditions, momentary distractions, and improper communications, which may result in failures.
- *Classification 4: Human ability limitations*—An example of human ability limitations is the limited capacity of short-term memory in the internal control mechanism.

A study highlighted the following seven causal factors, in order of greatest to least frequency of occurrence, for critical incidents and reported events concerning maintenance error in power plants [49,50]:

1. *Faulty procedures:* Faulty procedures are the most often appearing causal factor in the mishaps reported. It includes items such as wrong procedures, lack of adherence to a stated procedure, incompleteness, and lack of specificity. An example of faulty procedures is "due to poor judgement and not following stated guidelines properly, a ground was left on a circuit breaker. When the equipment was put back into service, the circuit breaker blew up and caused extensive property damage." In this situation, the proper procedure would have required clearing the ground before returning the circuit breaker to service.

2. *Problems in clearing and tagging equipment for maintenance:* Problems in clearing and tagging equipment for maintenance are a causal factor that is the second most frequent in reported cases where potentially serious accidents/serious accidents could be attributed to a failure/error associated with the equipment clearance process.

3. *Shortcomings in equipment design:* Shortcomings in equipment design are a causal factor that is the third most frequent for near accidents/accidents involving equipment design-associated problems. The factor includes items such as equipment incorrectly installed from the outset, poorly designed and inherently unreliable parts, the equipment not designed with proper mechanical safeguards for preventing the substitution of an incorrect part for the proper replacement part, and parts placed in inaccessible locations.

4. *Problems in moving people or equipment:* Problems in moving people or equipment are a causal factor that is the fourth most frequent. These problems basically stem from the inability to use appropriate vehicular aids in moving heavy units of equipment or poor lifting capability.

5. *Poor training, poor unit and equipment identification, and problems in facility design:* Poor training, poor unit and equipment identification, and problems in facility design are three causal factors that are the fifth most frequent. The factor *poor training* is basically concerned with the repair personnel's unfamiliarity with the task or their lack

of awareness of the system characteristics and inherent dangers associated with the task at hand.

The factor *poor unit and equipment identification* is the cause of an unexpected occurrence of a high number of accidents, and frequently, the problem is confusion between two identical parts/units and sometimes incorrect identification of potential hazards.

The factor *problems in facility design* can contribute to accidents. An example of these problems is insufficient clearances for repair personnel, equipment, or transportation aids in the performance of maintenance tasks.

6. *Poor work practices:* This causal factor is the sixth most frequent. Two examples of poor work practices are not taking the time to erect a scaffold so that an item in midair can be safely accessed and not waiting for operators to complete the switching and tagging tasks essential for disabling the systems/units requiring attention.

7. *Adverse environmental factors and mistakes by maintenance personnel:* Adverse environmental factors and mistakes by maintenance personnel are two causal factors that are the seventh (or the least) frequent.

The factor *adverse environmental factors* includes items such as the encouragement of haste by the need to minimize stay time in, say, radioactive environments and the need to wear protective devices and garments in threatening environments that, in turn, restrict the individual's movement capabilities and visual field.

The factor *mistakes by maintenance personnel* is a small fraction of those errors that would be quite difficult to anticipate and *design-out* of power generation plants.

Additional information on the seven causal factors previously mentioned is available in the study by Seminara and Parsons [50].

In the 1990s, the Electric Power Research Institute in the United States and the Central Research Institute of Electric Power Industry in Japan jointly conducted a study to identify critical maintenance tasks and for developing, implementing, and evaluating interventions that have high potential for reducing the occurrence of human errors or increasing maintenance-related productivity in NPPs. The study highlighted the following five most susceptible maintenance tasks to human error in power generation [51]:

- Overhaul motor-operated valve actuator
- Replace reactor coolant pump seals
- Overhaul main feed water pump
- Test reactor protection system
- Overhaul mainstream isolation valves

It simply means that careful attention is essential in performing such tasks for eliminating or minimizing the occurrence of human errors.

10.13 Guidelines to Reduce and Prevent Human Error in Power Generation Maintenance

Over the years, various guidelines have been proposed for reducing and preventing the occurrence of human error in power generation maintenance-related activities. Four of these guidelines are presented in the following [7].

- *Guideline 1: Develop appropriate work safety checklists for maintenance personnel.* Guideline 1 means that maintenance personnel should be provided with appropriate safety checklists, which can be used for determining the occurrence of human error and the factors that may affect their actions before or after the performance of maintenance-related tasks.

- *Guideline 2: Ameliorate design-related deficiencies.* As shortcomings in design can reduce attention to the activities and may even induce human error, Guideline 2 calls for overcoming shortcomings or deficiencies in areas such as work environment, labeling plant layout, and coding.

- *Guideline 3: Revise training programs for all concerned maintenance personnel.* Guideline 3 basically means that training programs for maintenance personnel should be revised in accordance with the characteristics and frequency of each extrinsic cause's occurrence.

- *Guideline 4: Perform administrative-related policies more thoroughly.* Guideline 4 basically means motivating all involved maintenance personnel appropriately to comply with prescribed quality control-related procedures.

Additional information on the previously mentioned guidelines is available in the study by Wu and Hwang [7].

10.14 Power Plant Maintenance Error Analysis Methods

There are many methods and models that can be used for performing maintenance error analysis in power generation. Two such methods/models are presented in the following.

10.14.1 Maintenance Personnel Performance Simulation Model

The maintenance personnel performance simulation (MAPPS) model was developed by the Oak Ridge National Laboratory, to provide estimates of NPP maintenance workforce performance measures, and it is a computerized, stochastic, task-oriented human behavioral model [51]. The development of this model was sponsored by the US Nuclear Regulatory Commission, and the basic objective for its development was the need for and the lack of human reliability-related data bank concerning NPP maintenance-related tasks, for use in conducting probabilistic risk assessment studies.

The measures of performance estimated by the MAPPS model include the probability of an undetected error, the probability of successfully completing the task of interest, the maintenance team stress profiles during task execution, the task duration time, and the most and least likely error-prone identification of subelements. Needless to say, the MAPPS model is a quite powerful tool to estimate important maintenance-related parameters and its flexibility permits it to be useful for various applications concerning NPP maintenance-related activity.

Additional information on this model is available in the study by Knee [52].

10.14.2 Fault Tree Analysis

FTA is a widely used method, particularly in the power generation industrial sector, for performing various types of reliability analysis [52,53]. The method is described in detail in Chapter 4. The following example demonstrates its application to the performance of maintenance error analysis in the area of power generation.

Example 10.4

Assume that a system used in a power plant can fail due to a maintenance error caused by three factors: use of deficient maintenance manuals, carelessness, or poor work environment. Two major factors for carelessness are time constraints and poor training. Similarly, two factors for poor work environment are distractions or inadequate lighting.

Develop a fault tree for the top event *Power plant system failure due to a maintenance error* by using fault tree symbols given in Chapter 4.

A fault tree for the example is shown in Figure 10.7.

Example 10.5

Assume that the probabilities of occurrence of events E_1, E_2, E_3, E_4, and E_5 shown in Figure 10.7 are 0.01, 0.02, 0.03, 0.04, and 0.05, respectively. For independent events, calculate the probabilities of occurrence of the top event T (i.e., power plant system failure due to a maintenance error) and intermediate events I_1 (i.e., poor work environment) and I_2 (i.e., carelessness) also shown in Figure 10.7.

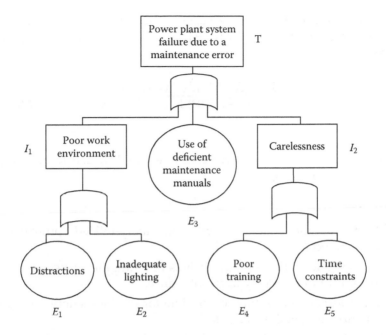

FIGURE 10.7
Fault tree for Example 10.4.

Using Chapter 4, the studies by Dhillon [36] and Dhillon and Singh [53], and the specified data values, we obtain the values of I_1, I_2, and T as follows: The probability of occurrence of event I_1 is given by

$$P(I_1) = P(E_1) + P(E_2) - P(E_1)P(E_2)$$
$$= 0.01 + 0.02 - (0.01)(0.02) \qquad (10.21)$$
$$= 0.0298,$$

where
$P(I_1)$, $P(E_1)$, and $P(E_2)$ are the probabilities of the occurrence of events I_1, E_1, and E_2, respectively.

The probability of occurrence of event I_2 is given by

$$P(I_2) = P(E_4) + P(E_5) - P(E_4)P(E_5)$$
$$= 0.04 + 0.05 - (0.04)(0.05) \qquad (10.22)$$
$$= 0.088,$$

where
$P(I_2)$, $P(E_4)$, and $P(E_5)$ are the probabilities of occurrence of events I_2, E_4, and E_5, respectively.

With the aid of the previously mentioned calculated and given values, Chapter 4, and the studies by Knee [52] and Dhillon and Singh [53], we get

$$P(T) = 1 - [1 - P(I_1)][1 - P(E_3)][1 - P(I_2)]$$
$$= 1 - (1 - 0.0298)(1 - 0.03)(1 - 0.088) \qquad (10.23)$$
$$= 0.1417.$$

Thus, the probabilities of occurrence of events T, I_1, and I_2 are 0.1417, 0.0298, and 0.088, respectively.

PROBLEMS

1. What are the main reasons for safety-related problems in maintenance?
2. Discuss at least 10 useful guidelines for equipment/system designers for improving safety in maintenance.
3. List at least seven factors responsible for dubious safety reputation in performing maintenance tasks.
4. Prove Equations 10.4 through 10.6 by using Equations 10.1 through 10.3.
5. List at least seven facts, figures, and examples concerning human error in aviation maintenance.
6. What are the main reasons, directly or indirectly, for the occurrence of human error in aviation maintenance?
7. List at least eight common human errors in aircraft maintenance tasks.
8. What are the methods that can be used for performing aircraft maintenance error analysis?
9. Write an essay on human error in power plant maintenance.
10. What are the main causes for the occurrence of human error in power plant maintenance?

References

1. *Accident Facts*, National Safety Council, Chicago, 1999.
2. Shepherd, W.T., The FAA human factors program in aircraft maintenance and inspection, *Proceedings of 5th Federal Aviation Administration (FAA) Meeting on Human Factors Issues in Aircraft Maintenance and Inspection*, 1991, June, pp. 1–5.

3. Hobbs, A., Williamson, A., Human factors in airline maintenance, *Proceedings of the Conference on Applied Psychology*, 1995, pp. 384–393.
4. Shepherd, W.T., Johnson, W.B., Drury, C.G., Berninger, D., *Human Factors in Aviation Maintenance Phase One: Progress Report*, Report No. AM-91/16, Office of Aviation Medicine, Federal Aviation Administration (FAA), Washington, DC, November 1991.
5. *Human Factors in Aircraft Maintenance and Inspection*, Report No. CAP 718, Safety Regulation Group, Civil Aviation Authority, London, 2002. Available from the Stationary Office, Norwich.
6. *Human Factors Guidelines for Aircraft Manual*, Report No. DOC 9824-AN/450, International Civil Aviation Organization, Montreal, 2003.
7. Wu, T.M., Hwang, S.L., Maintenance error reduction strategies in nuclear power plants, using root cause analysis, *Applied Ergonomics*, Vol. 20, No. 2, 1989, pp. 115–121.
8. Speaker, D.M., Voska, K.J., Luckas, W.J., *Identification and Analysis of Human Errors Underlying Electrical/Electronic Component Related Events*, Report No. NUREG/CR-2987, Nuclear Power Plant Operations, US Nuclear Regulatory Commission, Washington, DC, 1983.
9. Dhillon, B.S., *Engineering Maintenance: A Modern Approach*, CRC Press, Boca Raton, FL, 2002.
10. Dhillon, B.S., *Engineering Safety: Fundamentals, Techniques, and Applications*, World Scientific Publishing, NJ, 2003.
11. Latino, C.J., *Hidden Treasure: Eliminating Chronic Failures Can Cut Maintenance Costs up to 60%*, Report, Reliability Center, Hopewell, VA, 1999.
12. *Human Factors in Airline Maintenance: A Study of Incident Reports, Bureau of Air Safety Investigation*, Department of Transport and Regional Development, Canberra, 1997.
13. Russell, P.D., Management strategies for accident prevention, *Air Asia*, Vol. 6, 1994, pp. 31–41.
14. Australian Transport Safety Bureau, *ATSB Survey of Licensed Aircraft Maintenance Engineers in Australia*, Report No. ISBN 0642274738, Australian Transport Safety Bureau, Department of Transport and Regional Services, Canberra, 2001.
15. Gero, D., *Aviation Disasters*, Published by Patrick Stephens, Sparkford, 1993.
16. Goetsch, D.L., *Occupational Safety and Health*, Prentice Hall, Englewood Cliffs, NJ, 1996.
17. Christensen, J.M., Howard, J.M., Field experience in maintenance, in *Human Detection and Diagnosis of System Failures*, edited by J. Rasmussen and W.B., Rouse, Plenum Press, New York, 1981, pp. 111–133.
18. *Joint Fleet Maintenance Manual*, Vol. 5, Quality Assurance, Submarine Maintenance Engineering, US Navy, Portsmouth, NH.
19. Ministry of Transportation, *Report on the Accident to BAC 1-11*, Report No. 1-92, Air Accident Investigation Branch, Ministry of Transportation, London, 1992.
20. Stoneham, D., *The Maintenance Management and Technology Handbook*, Elsevier Science, Amsterdam, 1998.
21. Hammer, W., *Product Safety Management and Engineering*, Prentice Hall, Englewood Cliffs, NJ, 1980.

22. Dhillon, B.S., *Design Reliability: Fundamentals and Applications*, CRC Press, Boca Raton, FL, 1999.
23. Marx, D.A., *Learning From our Mistakes: A Review of Maintenance Error Investigation and Analysis Systems (with recommendations to the FAA)*, Federal Aviation Administration (FAA), Washington, DC, 1998, January.
24. Phillips, E.H., Focus on accident prevention key to future airline safety, *Aviation Week and Space Technology*, Issue No. 5, 1994, pp. 52–53.
25. Kraus, D.C., Gramopadhys, A.K., Effect of team training on aircraft maintenance technicians: Computer-based training versus instructor-based training, *International Journal of Industrial Ergonomics*, Vol. 27, 2001, pp. 141–157.
26. Marx, D.A., Graeber, R.C., Human error in maintenance, in *Aviation Psychology in Practice*, edited by N. Johnston, N. McDonald, and R. Fuller, Ashgate Publishing, London, 1994, pp. 87–104.
27. Christensen, J.M., Howard, J.M., Field experience in maintenance, in *Human Detection and Diagnosis of System Failures*, edited by J. Rasmussen and W.B. Rouse, Plenum Press, New York, 1981, pp. 111–133.
28. Latorella, K.A., Prabhu, P.V., A review of human error in aviation maintenance and inspection, *International Journal of Industrial Ergonomics*, Vol. 26, 2000, pp. 133–161.
29. Graeber, R.C., Max, D.A., Reducing human error in aircraft maintenance operations, *Proceedings of the 46th Annual International Safety Seminar*, 1993, pp. 147–160.
30. *Human Factors in Airline Maintenance: A Study of Incident Reports*, Report No. 2-97, Bureau of Air Safety Investigation (BASI), Department of Transport and Regional Development, Canberra, 1997.
31. Russell, P.D., Management strategies for accident prevention, *Air Asia*, Vol. 6, 1994, pp. 31–41.
32. Wenner, C.A., Drury, C.G., Analyzing human error in aircraft ground damage incidents, *International Journal of Industrial Ergonomics*, Vol. 26, 2000, pp. 177–199.
33. Prabhu, P., Drury, C.G., A framework for the design of the aircraft inspection information environment, *Proceedings of the 7th FAA Meeting on Human Factors Issues in Aircraft Maintenance and Inspection*, 1992, pp. 54–60.
34. Dhillon, B.S., *Human Reliability and Error in Transportation Systems*, Springer-Verlag, London, 2007.
35. *Investigation of Human Factors in Accidents and Incidents*, Report No. 93-1, International Civil Aviation Organization, Montreal, 1993.
36. Dhillon, B.S., *Human Reliability: With Human Factors*, Pergamon Press, New York, 1986.
37. Dhillon, B.S., *Human Reliability, Error, and Human Factors in Engineering Maintenance*, CRC Press, Boca Raton, FL, 2009.
38. Allen, J.P., Rankin, W.L., A summary of the use and impact of the Maintenance Error Decision Aid (MEDA) on the commercial aviation industry, *Proceedings of the 48th Annual International Air Safety Seminar*, 1995, pp. 359–369.
39. Swain, A.D., An error-cause removal program for industry, *Human Factors*, Vol. 12, 1973, pp. 207–221.
40. Hasegawa, T., Kameda, A., Analysis and evaluation of human error events in nuclear power plants, Presented at the Meeting of the IAEA's CRP on "Collection and Classification of Human Reliability Data for Use in Probabilistic Safety Assessments," Available from the Institute of Human Factors, Nuclear Power Engineering Corporation, Tokyo, 1998, May.

41. Daniels, R.W., The formula for improved plant maintainability must include human factors, *Proceedings of the IEEE Conference on Human Factors and Nuclear Safety*, 1985, pp. 242–244.
42. Reason, J., Human factors in nuclear power generation: A systems perspective, *Nuclear Europe Worldscan*, Vol. 17, Nos. 5–6, 1997, pp. 35–36.
43. *Nuclear Power Plant Operating Experience*, from the IAEA/NEA Incident Reporting System 1996–1999, Organization for Economic Co-operation and Development (OECD), Paris, 2000.
44. Pyy, P., Laakso, K., Reiman, L., A study of human errors related to NPP maintenance activities, *Proceedings of the IEEE 6th Annual Human Factors Meeting*, 1997, pp. 12.23–12.28.
45. Pyy, P., An analysis of maintenance failures at a nuclear power plant, *Reliability Engineering and System Safety*, Vol. 72, 2001, pp. 293–302.
46. *Detroit News*, The UAW and the Rouge explosion: A pat on the head, Detroit, MI, 1999, February 6, p. 6.
47. White, J., New revelations expose company-union complexity in fatal blast at US Ford Plant. World Socialist Web Site. Available online at www.wsws.org/articles/2000/feb2000/ford-fo4.shtml.
48. *Miami Herald*, Maintenance error a factor in blackouts, Miami, FL, 1989, December 29, p. 4.
49. Seminara, J.L., Parsons, S.O., Human factors engineering and power plant maintenance, *Maintenance Management International*, Vol. 6, 1985, pp. 33–71.
50. Seminara, J.L., Parsons, S.O., *Human Factors Review of Power Plant Maintainability*, Report No. NP-1567 (Research project 1136), Electric Power Research Institute, Palo Alto, CA, 1981.
51. Isoda, H., Yasutake, J.Y., Human factors interventions to reduce human errors and improve productivity in maintenance tasks, *Proceedings of the International Conference on Design and Safety of Advanced Nuclear Power Plants*, 1992, pp. 34.4.1–34.4.6.
52. Knee, H.E., The maintenance personnel performance simulation (MAPPS) model: A human reliability analysis tool, *Proceedings of the International Conference on Nuclear Power Plant Aging, Availability Factor, and Reliability Analysis*, 1985, pp. 77–80.
53. Dhillon, B.S., Singh, C., *Engineering Reliability: New Techniques and Applications*, Wiley, New York, 1981.

11

Mathematical Models for Performing Engineering System Reliability, Safety, and Maintenance Analysis

11.1 Introduction

In the area of engineering, mathematical modeling is a commonly used approach to performing various types of analysis. In this case, an engineering system's parts are represented as idealized elements assumed to have the representative characteristics of real-life parts, whose behavior is possible to be described by equations. However, it is to be noted that a mathematical model's degree of realism very much depends on the type of assumptions imposed upon it.

Over the years, many mathematical models have been developed to perform engineering system reliability, safety, and maintenance analysis. Many of these models were developed using the Markov method [1–4]. Although the effectiveness of such models can considerably vary from one application area to another, some of them are being used quite successfully for performing engineering system reliability, safety, and maintenance analysis.

This chapter presents a number of mathematical models considered useful for performing various types of engineering systems reliability, safety, and maintenance analysis.

11.2 Model I

Model I represents an engineering system that can be in any one of the three states: system operating normally in the field, system failed in the field, and failed system in the workshop for repair. More clearly, the failed system is always taken to the workshop for repair. The repaired system is put back to its normal operating state. A typical example of such system is a motor vehicle.

The system state-space diagram is shown in Figure 11.1. The numerals in the box, the circle, and the diamond denote system states. The model is subjected to the following assumptions:

- System failures occur independently.
- System failure, repair, and towing rates are constant.
- The repaired system is as good as new.

The following symbols are associated with the state-space diagram shown in Figure 11.1 and its associated equations:

j is the jth state of the engineering system, where $j = 0$ (system operating normally in the field), $j = 1$ (system failed in the field), $j = 2$ (failed system in the repair workshop).

λ_f is the system constant failure rate.

θ is the system constant repair rate.

λ_t is the system constant towing rate from state 1 to state 2.

$P_j(t)$ is the probability that the engineering system is in state j at time t, for $j = 0, 1, 2$.

Using the Markov method described in Chapter 4, we write down the following equations for Figure 11.1 state-space diagram [5]:

$$\frac{dP_0(t)}{dt} + \lambda_f P_0(t) = \theta P_2(t), \tag{11.1}$$

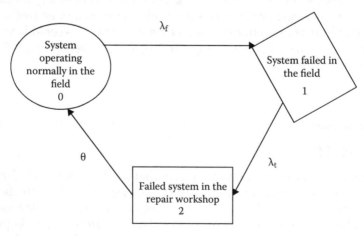

FIGURE 11.1
System state-space diagram.

$$\frac{dP_1(t)}{dt} + \lambda_t P_1(t) = \lambda_f P_0(t), \tag{11.2}$$

$$\frac{dP_2(t)}{dt} + \theta P_2(t) = \lambda_t P_1(t). \tag{11.3}$$

At time $t = 0$, $P_0(0) = 1$, $P_1(0) = 0$, and $P_2(0) = 0$.

By solving Equations 11.1 through 11.3, we obtain the following steady-state probability equations [5]:

$$P_0 = \left(1 + \frac{\lambda_f}{\lambda_t} + \frac{\lambda_f}{\theta}\right)^{-1}, \tag{11.4}$$

$$P_1 = \left(\frac{\lambda_f}{\lambda_t}\right) P_0, \tag{11.5}$$

$$P_2 = \left(\frac{\lambda_f}{\theta}\right) P_0, \tag{11.6}$$

where

P_0, P_1, and P_2 are the steady-state probabilities of the engineering system being in states 0, 1, and 2, respectively.

The engineering system steady-state availability is given by

$$AV_{es} = P_0, \tag{11.7}$$

where

AV_{es} is the engineering system steady-state availability.

By setting $\theta = 0$ in Equations 11.1 through 11.3 and then solving the resulting equations, we get

$$R_{es}(t) = P_0(t) = e^{-\lambda_f t}, \tag{11.8}$$

where

$R_{es}(t)$ is the engineering system reliability at time t.

The engineering system MTTF is given by Dhillon [1] as

$$MTTF_{es} = \int_0^\infty R_{es}(t)\,dt$$

$$= \int_0^\infty e^{-\lambda_f t}\,dt \qquad (11.9)$$

$$= \frac{1}{\lambda_f},$$

where
$MTTF_{es}$ is the engineering system MTTF.

Example 11.1

Assume that the constant failure rate of a three-state engineering system is 0.002 failures/hour. Calculate the engineering system MTTF and its reliability during an 8-hour mission.

By substituting the given data value into Equation 11.9, we obtain

$$MTTF_{es} = \frac{1}{0.002} = 500 \text{ hours.}$$

Using the given data values in Equation 11.8 yields

$$R_{es}(8) = e^{-(0.002)(8)}$$

$$= 0.9841.$$

Thus, the engineering system MTTF and reliability are 500 hours and 0.9841, respectively.

11.3 Model II

Model II represents an engineering system operating in a fluctuating environment (e.g., normal and stormy weather). The system can fail operating either in a normal or an abnormal environment. The failed system is repaired back to both its operating states. The engineering system state-space diagram is shown in Figure 11.2. The numerals in the circles and the box denote system states.

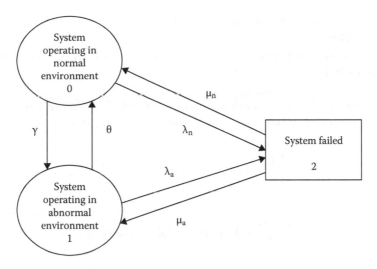

FIGURE 11.2
System state-space diagram.

The model is subjected to the following assumptions:

- System failures occur independently.
- Fluctuating environment transition rates (i.e., from normal environment state to abnormal environment state and vice versa) are constant.
- System failure and repair rates are constant.
- The repaired system is as good as new.

The following symbols are associated with the state-space diagram shown in Figure 11.2 and its associated equations:

j is the jth state of the engineering system, where $j = 0$ (system operating in normal environment), $j = 1$ (system operating in abnormal environment), $j = 2$ (system failed).

γ is the constant changeover rate of the environment from state 0 to state 1.

θ is the constant changeover rate of the environment from state 1 to state 0.

λ_n is the system constant failure rate from the normal environment operating state.

λ_a is the system constant failure rate from the abnormal environment operating state.

μ_n is the system constant repair rate (normal environment) from state 2 to state 0.

μ_a is the system constant repair rate (abnormal environment) from state 2 to state 1.

$P_j(t)$ is the probability that the engineering system is in state j at time t; for $j = 0, 1, 2$.

Using the Markov method described in Chapter 4, we write down the following equations for Figure 11.2 state-space diagram [6]:

$$\frac{dP_0(t)}{dt} + (\gamma + \lambda_n)P_0(t) = \theta P_1(t) + \mu_n P_2(t), \tag{11.10}$$

$$\frac{dP_1(t)}{dt} + (\theta + \lambda_a)P_1(t) = \gamma P_0(t) + \mu_a P_2(t), \tag{11.11}$$

$$\frac{dP_2(t)}{dt} + (\mu_u + \mu_a)P_2(t) = \lambda_n P_0(t) + \lambda_a P_1(t). \tag{11.12}$$

At time $t = 0$, $P_0(0) = 1$, $P_1(0) = 0$, and $P_2(0) = 0$.
By solving Equations 11.10 through 11.12, we get the following steady-state probability equations [6]:

$$P_0 = B_1 / X_1 X_2, \tag{11.13}$$

where

$$B_1 = \mu_n\theta + \lambda_a\mu_n + \theta\mu_a, \tag{11.14}$$

$$X_1, X_2 = \frac{-Z \pm \left[Z^2 - 4\left(B_1 + B_2 + B_3\right)\right]^{1/2}}{2}, \tag{11.15}$$

$$B_2 = \gamma\lambda_a + \theta\mu_n + \lambda_n\lambda_a, \tag{11.16}$$

$$B_3 = \gamma\mu_n + \gamma\mu_a + \lambda_n\mu_a, \tag{11.17}$$

$$Z = \theta + \mu_n + \mu_a + \gamma + \lambda_n + \lambda_a, \tag{11.18}$$

$$P_1 = B_3/X_1X_2, \tag{11.19}$$

$$P_2 = B_2/X_1X_2, \tag{11.20}$$

where

P_0, P_1, and P_2 are the steady-state probabilities of the engineering system being in states 0, 1, and 2, respectively.

The engineering system steady-state availability in both types of environment is given by

$$AV_{ss} = P_0 + P_1, \tag{11.21}$$

where

AV_{ss} is the engineering system steady-state availability in both types of environment.

By setting $\mu_n = \mu_a = 0$ in Equations 11.10 through 11.12 and then solving the resulting equations and using the studies by Dhillon [1,6], we obtain

$$MTTF_s = \lim_{s \to 0} R_s(s) = \lim_{s \to 0}\{P_0(s) + P_1(s)\}$$
$$= \frac{\lambda_a + \gamma + \theta}{(\lambda_n + \gamma)(\lambda_a + \theta) - \lambda\theta}, \tag{11.22}$$

where

$MTTF_s$ is the engineering system MTTF.
s is the Laplace transform variable.
$R_s(s)$ is the Laplace transform of the engineering system reliability.
$P_0(s)$ is the Laplace transform of the probability that the engineering system is in state 0.
$P_1(s)$ is the Laplace transform of the probability that the engineering system is in state 1.

Example 11.2

Assume that in Equation 11.22, we have the following given data values:

$$\lambda_a = 0.002 \text{ failures/hour,}$$

$$\lambda_n = 0.005 \text{ failures/hour,}$$

$$\theta = 0.003 \text{ transitions/hour,}$$

$$\gamma = 0.001 \text{ transitions/hour.}$$

Calculate the engineering system MTTF.
By inserting the specified data values into Equation 11.22, we obtain

$$MTTF_s = \frac{0.002 + 0.001 + 0.003}{\left[(0.005 + 0.001)(0.002 + 0.003)\right] - \left[(0.001)(0.003)\right]}$$

$$= 222.22 \text{ hours.}$$

Thus, the engineering system MTTF is 222.22 hours.

11.4 Model III

Model III represents an engineering system that can be in any one of the three states: operating normally, operating unsafely, and failed. The system is repaired from unsafe working state and fully failed state. The system state-space diagram is shown in Figure 11.3. The numerals in the box, the circle, and the diamond denote system states.

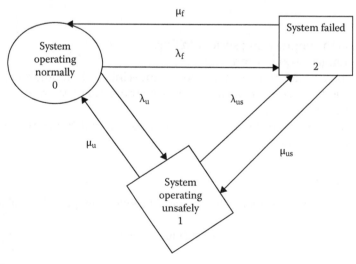

FIGURE 11.3
System state-space diagram.

The model is subjected to the following assumptions:

- System failures occur independently.
- System failure and repair rates are constant.
- The repaired system is as good as new.

The following symbols are associated with the state-space diagram shown in Figure 11.3 and its associated equations:

j is the jth state of the engineering system; $j = 0$ means the system is operating normally; $j = 1$ means the system is operating unsafely; $j = 2$ means the system failed.

λ_j is the system jth constant failure rate; $j = f$ means from state 0 to state 2; $j = u$ means from state 0 to state 1; $j = us$ means from state 1 to state 2.

μ_j is the system jth constant repair rate; $j = u$ means from state 1 to state 0; $j = f$ means from state 2 to state 0; $j = us$ means from state 2 to state 1.

$P_j(t)$ is the probability that the engineering system is in state j at time t, for $j = 0, 1, 2$.

Using the Markov method described in Chapter 4, we write down the following equations for Figure 11.3 state-space diagram [3,7]:

$$\frac{dP_0(t)}{dt} + \left(\lambda_u + \lambda_f\right)P_0(t) = \mu_u P_1(t) + \mu_f P_2(t), \qquad (11.23)$$

$$\frac{dP_1(t)}{dt} + \left(\mu_u + \mu_{us}\right)P_1(t) = \mu_{us}P_2(t) + \lambda_u P_0(t), \qquad (11.24)$$

$$\frac{dP_2(t)}{dt} + \left(\mu_f + \mu_{us}\right)P_2(t) = \lambda_{us}P_1(t) + \lambda_f P_0(t). \qquad (11.25)$$

At time $t = 0$, $P_0(0) = 1$ and $P_1(0) = P_2(0) = 0$.

For a very large t, by solving Equations 11.23 through 11.25, we get the following steady-state probability equations:

$$P_0 = \frac{\left(\mu_f + \mu_{us}\right)\left(\mu_u + \lambda_{us}\right) - \lambda_{us}\mu_{us}}{C}, \qquad (11.26)$$

where

$$C = (\mu_f + \mu_{us})(\mu_u + \lambda_u + \lambda_{us}) + \lambda_f(\mu_u + \lambda_{us}) + \lambda_f\mu_{us} + \lambda_u\lambda_{us} - \lambda_{us}\mu_{us}$$

$$P_1 = \frac{\lambda_u(\mu_f + \mu_{us}) + \lambda_f\mu_{us}}{C}, \tag{11.27}$$

$$P_2 = \frac{\lambda_f\lambda_{us} + \lambda_f(\mu_u + \lambda_{us})}{C}, \tag{11.28}$$

where

P_0, P_1, and P_2 are the steady-state probabilities of the engineering system being in states 0, 1, and 2, respectively.

It is to be noted that the steady-state probability of the system operating unsafely is given in Equation 11.27.

By setting $\mu_f = \mu_{us} = 0$ in Equations 11.23 through 11.25, we get the following equation for the system reliability (i.e., the system operating normally and unsafely probability):

$$R_{es}(t) = P_0(t) + P_1(t)$$
$$= (C_1 + D_1)e^{x_1 t} + \left[(C_2 + D_2)e^{x_2 t}\right], \tag{11.29}$$

where

$R_{es}(t)$ is the engineering system reliability at time t.

$x_1 = -N_1 + (N_1^2 - 4N_2)^{1/2}$.

$x_2 = -N_1 - (N_1^2 - 4N_2)^{1/2}$.

$N_1 = \mu_u + \lambda_f + \lambda_u + \lambda_{us}$.

$N_2 = \lambda_f\mu_u + \lambda_f\lambda_{us} + \lambda_u\lambda_{us}$.

$C_1 = \dfrac{x_1 + \mu_u + \lambda_{us}}{(x_1 - x_2)}$.

$C_2 = \dfrac{x_2 + \mu_u + \lambda_{us}}{(x_2 - x_1)}$.

$D_1 = \dfrac{\lambda_u}{(x_1 - x_2)}$.

$D_2 = \dfrac{\lambda_u}{(x_2 - x_1)}$.

By integrating Equation 11.29 over the time interval $[0, \infty]$, we obtain the following equation for the engineering system MTTF ($MTTF_{es}$):

$$MTTF_{es} = \int_0^\infty R_{es}(t)dt$$

$$= -\left[\frac{(C_1 + D_1)}{x_1} + \frac{(C_2 + D_2)}{x_2}\right].$$

(11.30)

Example 11.3

Assume that an engineering system can be either operating normally, operating unsafely, or failed. Its failure rates from the normal operating state to the unsafe operating state, the unsafe operating state to the fully failed state, and the normal operating state to the fully failed state are 0.006, 0.003, and 0.07 failures/hour, respectively. Similarly, the engineering system repair rates from the fully failed state to the normal operating state, the unsafe operating state to the normal operating state, and the fully failed state to the unsafe operating state are 0.08, 0.009, and 0.004 repairs/hour, respectively.

Calculate the probability of the engineering system being in unsafe operating state during a very large mission period, if the failure and repair rates associated with the system are constant.

By inserting the given data values into Equation 11.27, we get

$$P_1 = \frac{(0.006)(0.08 + 0.004) + (0.07)(0.004)}{C}$$

$$C = (0.08 + 0.004)(0.009 + 0.006 + 0.003) + (0.07)(0.009 + 0.003)$$
$$+ (0.07)(0.004) + (0.006)(0.003) - (0.003)(0.004)$$
$$= 0.2971.$$

Thus, the probability of the engineering system being in unsafe operating state during a very large mission period is 0.2971.

11.5 Model IV

This mathematical model represents an engineering system having three states: operating normally, failed unsafely, and failed safely. The failed system

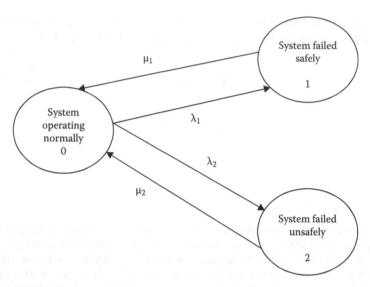

FIGURE 11.4
System state-space diagram.

is repaired, and the system state-space diagram is shown in Figure 11.4. The numerals in the circles denote system states.

The model is subjected to the following assumptions:

- System failures occur independently.
- The system can fail either safely or unsafely.
- The system failure and repair rates are constant.
- The repaired system is as good as new.

The following symbols are associated with the state-space diagram shown in Figure 11.4 and its associated equations:

j is the jth state of the engineering system: $j = 0$ means the system is operating normally; $j = 1$ means the system failed safely; $j = 2$ means the system failed unsafely.

λ_j is the system jth constant failure rate; $j = 1$ means safe; $j = 2$ means unsafe.

μ_j is the failed system jth constant repair rate; $j = 1$ means from safe failed state; $j = 2$ means from unsafe failed state.

$P_j(t)$ is the probability that the engineering system is in state j at time t; for $j = 0, 1, 2$.

Using the Markov method described in Chapter 4, we write down the following equations for Figure 11.4 state-space diagram [1,3,8]:

$$\frac{dP_0(t)}{dt} + (\lambda_1 + \lambda_2)P_0(t) = \mu_1 P_1(t) + \mu_2 P_2(t), \tag{11.31}$$

$$\frac{dP_1(t)}{dt} + \mu_1 P_1(t) = \lambda_1 P_0(t), \tag{11.32}$$

$$\frac{dP_2(t)}{dt} + \mu_2 P_2(t) = \lambda_2 P_0(t). \tag{11.33}$$

At time $t = 0$, $P_0(0) = 1$ and $P_1(0) = P_2(0) = 0$.

By solving Equations 11.31 through 11.33, we obtain

$$P_0(t) = \frac{\mu_1 \mu_2}{Z_1 Z_2} + \left[\frac{(Z_1 + \mu_1)(Z_1 + \mu_2)}{Z_1(Z_1 - Z_2)}\right]e^{Z_1 t} - \left[\frac{(Z_2 + \mu_1)(Z_2 + \mu_2)}{Z_2(Z_1 - Z_2)}\right], \tag{11.34}$$

where

$$Z_1, Z_2 = \frac{-B \pm \left[B^2 - 4(\mu_1 \mu_2 + \lambda_1 \mu_2 + \lambda_2 \mu_2)\right]^{1/2}}{2},$$

$$B = \mu_1 + \mu_2 + \lambda_1 + \lambda_2,$$

$$Z_1 Z_2 = \mu_1 \mu_2 + \lambda_1 \mu_2 + \lambda_2 \mu_1,$$

$$Z_1 + Z_2 = -(\mu_1 + \mu_2 + \lambda_1 + \lambda_2),$$

$$P_1(t) = \frac{\lambda_1 \mu_2}{Z_1 Z_2} + \left[\frac{(\lambda_1 Z_1 + \lambda_1 \mu_2)}{Z_1(Z_1 - Z_2)}\right]e^{Z_1 t} - \left[\frac{(\mu_2 + Z_2)\lambda_1}{Z_2(Z_1 - Z_2)}\right]e^{Z_2 t}, \tag{11.35}$$

$$P_2(t) = \frac{\lambda_2 \mu_1}{Z_1 Z_2} + \left[\frac{(\lambda_2 Z_1 + \lambda_2 \mu_1)}{Z_1(Z_1 - Z_2)}\right]e^{Z_1 t} - \left[\frac{(\mu_1 + Z_2)\lambda_2}{Z_2(Z_1 - Z_2)}\right]e^{Z_2 t}. \tag{11.36}$$

Equations 11.36 and 11.35 give the probability of the engineering system failing unsafely and safely, respectively, when subjected to the repair process.

As time t becomes very large, the engineering system steady-state probability of failing unsafely using Equation 11.36 is

$$P_2 = \lim_{t \to \infty} P_2(t) = \frac{\lambda_2 \mu_1}{Z_1 Z_2}. \tag{11.37}$$

Similarly, as time t becomes very large, the engineering system steady-state probability of failing safely using Equation 11.35 is

$$P_1 = \lim_{t \to \infty} P_1(t) = \frac{\lambda_1 \mu_2}{Z_1 Z_2}. \tag{11.38}$$

By setting $\mu_1 = \mu_2 = 0$ in Equations 11.34 through 11.36 and then solving the resulting equations, we obtain

$$P_0(t) = e^{-(\lambda_1 + \lambda_2)t}, \tag{11.39}$$

$$P_1(t) = \frac{\lambda_1}{\lambda_2 + \lambda_1} \left[1 - e^{-(\lambda_1 + \lambda_2)t} \right], \tag{11.40}$$

$$P_2(t) = \frac{\lambda_2}{\lambda_2 + \lambda_1} \left[1 - e^{-(\lambda_1 + \lambda_2)t} \right]. \tag{11.41}$$

Equations 11.40 and 11.41 give the probability of the engineering system failing safely and unsafely at time t, respectively. In contrast, Equation 11.39 gives the engineering system reliability or the probability of success at time t.

By integrating Equation 11.39 over the time interval $[0, \infty]$, we obtain the following expression for the engineering system MTTF [1]:

$$\begin{aligned} MTTF_{es} &= \int_0^\infty P_0(t)(d)t \\ &= \int_0^\infty e^{-(\lambda_1 + \lambda_2)t} \, dt \\ &= \frac{1}{\lambda_2 + \lambda_2}, \end{aligned} \tag{11.42}$$

where
$MTTF_{es}$ is the engineering system MTTF.

Example 11.4

Assume that an engineering system can fail unsafely or safely, and its unsafe and safe constant failure rates are 0.002 and 0.009 failures/hour, respectively. Similarly, its unsafe and safe failure mode constant repair rates are 0.004 and 0.05 repairs/hour, respectively.

Calculate the probability of the engineering system being in unsafe failure mode during a very large mission period.

By inserting the specified data values into Equation 11.37, we get

$$P_2 = \frac{\lambda_2 \mu_1}{Z_1 Z_2} = \frac{(0.002)(0.05)}{(0.05)(0.004) + (0.009)(0.004) + (0.002)(0.05)}$$
$$= 0.2976$$

Thus, the probability of the engineering system being in unsafe operating state during a very large mission period is 0.2976.

11.6 Model V

This mathematical model represents a situation when the preventive maintenance is performed on the engineering system as well as it is repaired back upon failure. The state-space diagram shown in Figure 11.5 denotes such a situation. The numeral and the single letters in the box and the circles denote engineering system states.

The model is subjected to the following assumptions:

- System failures occur independently.
- The system failure, preventive maintenance, and repair rates are constant.
- The repaired system is as good as new.

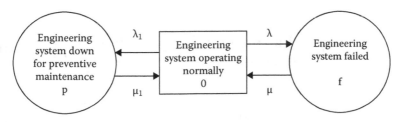

FIGURE 11.5
Engineering system state-space diagram.

The following symbols are associated with the state space diagram shown in Figure 11.5 and its associated equations:

j denotes the jth state of the engineering system: $j = 0$ (operating normally); $j = f$ (failed); $j = p$ (down for preventive maintenance).

λ denotes the engineering system constant failure rate.

μ denotes the engineering system constant repair rate.

λ_1 denotes the engineering system constant preventive maintenance rate.

μ_1 denotes the engineering system constant preventive maintenance accomplishment rate.

$P_j(t)$ denotes the probability that the engineering system is in state j at time t; for $j = 0, 1, 2$.

Using the Markov method described in Chapter 4, we write down the following equations for Figure 11.5 state-space diagram [3,8]:

$$\frac{dP_0(t)}{dt} + (\lambda_1 + \lambda)P_0(t) = \mu_1 P_p(t) + \mu P_f(t), \tag{11.43}$$

$$\frac{dP_p(t)}{dt} + \mu_1 P_p(t) = \lambda_1 P_0(t), \tag{11.44}$$

$$\frac{dP_f(t)}{dt} + \mu P_f(t) = \lambda P_0(t). \tag{11.45}$$

At time $t = 0$, $P_0(0) = 1$ and $P_f(0) = P_p(0) = 0$.

By solving Equations 11.43 through 11.45, we obtain

$$P_0(t) = \frac{\mu_1 \mu}{n_1 n_2} + \left[\frac{(n_1 + \mu_1)(n_1 + \mu)}{n_1(n_1 - n_2)}\right]e^{n_1 t} - \left[\frac{(n_1 + \mu_1)(n_2 + \mu)}{n_2(n_1 - n_2)}\right]e^{n_2 t}, \tag{11.46}$$

$$P_p(t) = \frac{\lambda_1 \mu}{n_1 n_2} \left[\frac{\lambda_1 n_1 + \lambda_1 \mu}{n_1(n_1 - n_2)}\right]e^{n_1 t} - \left[\frac{(\mu + n_2)\lambda_1}{n_2(n_1 - n_2)}\right]e^{n_2 t}, \tag{11.47}$$

$$P_f(t) = \frac{\lambda \mu_1}{n_1 n_2} + \left[\frac{\lambda n_1 + \lambda \mu_1}{n_1(n_1 - n_2)}\right]e^{n_1 t} - \left[\frac{(\mu_1 + n_2)\lambda}{n_2(n_1 - n_2)}\right]e^{n_2 t}, \tag{11.48}$$

where

$$n_1 n_2 = \mu_1 \mu + \lambda_1 \mu + \lambda \mu_1, \tag{11.49}$$

$$n_1 + n_2 = -(\mu_1 + \mu + \lambda_1 + \lambda). \tag{11.50}$$

The engineering system availability at time t is given by Equation 11.46. This availability expression is valid, if and only if n_1 and n_2 are negative.

As t becomes very large in Equation 11.46, the steady-state engineering system availability (A_{es}) is given by

$$A_{es} = \lim_{t \to \infty} P_0(t) = \frac{\mu \mu_1}{n_1 n_2}$$

$$= \frac{\mu \mu_1}{\mu_1 \mu + \lambda_1 \mu + \lambda \mu_1}. \tag{11.51}$$

Example 11.5

Assume that an engineering system either can be down for preventive maintenance or failed, and its constant preventive maintenance and failure rates are 0.008/hour and 0.002/hour, respectively. Similarly, its preventive maintenance and failure mode constant preventive maintenance accomplishment and repair rates are 0.06/hour and 0.04/hour, respectively.

Calculate the engineering system availability during a very large mission period.

By substituting the given data values into Equation 11.51, we get

$$A_{es} = \frac{(0.04)(0.06)}{(0.06)(0.04) + (0.008)(0.04) + (0.002)(0.06)}$$

$$= 0.8450.$$

Thus, the engineering system availability during a very large mission period is 0.8450.

11.7 Model VI

This mathematical model can be used to determine the optimum replacement interval of an engineering item where the increasing trend in the item maintenance cost is predictable [9,10]. For the increasing maintenance cost of an item, the annual total cost of that item is expressed by

$$C_{at} = C_n + \left[y(\alpha - 1)/2 \right] + \frac{AC}{\alpha}, \quad (11.52)$$

where
C_{at} is the item annual total cost.
C_n is the annual nonvarying maintenance and operating cost.
y is the annual increase in the maintenance cost.
α is the number of years of the item life.
AC is the acquisition cost (AC) of the item.

It is to be noted that in Equation 11.52, the interest rate is neglected.
To find the optimum value of α, we differentiate Equation 11.52 with respect to α, then set the resulting expression equal to zero. Thus, we have

$$\frac{dc_{at}}{d\alpha} = \frac{y}{2} - \frac{AC}{\alpha^2} = 0. \quad (11.53)$$

By rearranging Equation 11.53, we obtain

$$\alpha_0 = \left(\frac{2AC}{y} \right)^{1/2}, \quad (11.54)$$

where
α_0 is the optimum replacement period of the item.

Example 11.6

Find the optimum value of α in Equation 11.53 if the item/system AC is US$80,000, and the annual increase in the maintenance cost (i.e., y) is US$400 per year.
By substituting the given data values into Equation 11.54, we get

$$\alpha_0 = \left[\frac{2(80,000)}{400} \right]^{1/2}$$

$$= 20 \text{ years.}$$

Thus, the optimum value of α is 20 years (i.e., $\alpha_0 = 20$ *years*).

PROBLEMS

1. Prove Equations 11.4 through 11.6 by using Equations 11.1 through 11.3.

2. Prove Equation 11.22 by using Equations 11.10 through 11.12.

3. Assume that a system can be either working normally, working unsafely, or failed. Its failure rates from normal working state to unsafe working state, unsafe working state to fully failed state, and from normal working state to fully failed state are 0.005, 0.002, and 0.06 failures/hour, respectively. Similarly, the system repair rates from the fully failed state to normal working state, unsafe working state to normal working state, and fully failed state to unsafe working state are 0.07, 0.008, and 0.003 repairs/hour, respectively. Calculate the probability of the system being in unsafe working state during a very large mission period.

4. Prove Equation 11.38 by using Equation 11.35.

5. Assume that a system can fail safely or unsafely and its unsafe and safe constant failure rates are 0.001 and 0.008 failures/hour, respectively. Similarly, its unsafe and safe failure mode constant repair rates are 0.003 and 0.04 repairs/hour, respectively.

   Calculate the probability of the system being in unsafe failure mode during a very large mission period.

6. Prove that the sum of Equations 11.46 through 11.48 is equal to unity.

7. Obtain expressions for $P_0(t)$, $P_1(t)$, and $P_2(t)$ by using Equations 11.1 through 11.3.

8. Prove that the sum of Equations 11.13, 11.9, and 11.20 is equal to unity.

9. Prove Equation 11.30 by using Equation 11.29.

10. Prove that the sum of Equations 11.34 through 11.36 is equal to unity.

References

1. Dhillon, B.S., *Design Reliability: Fundamentals and Applications*, CRC Press, Boca Raton, FL, 1999.
2. Dhillon, B.S., *Applied Reliability and Quality: Fundamentals, Methods, and Procedures*, Springer, London, 2007.
3. Dhillon, B.S., *Engineering Safety: Fundamentals, Techniques, and Applications*, World Scientific Publishing, River Ridge, NJ, 2003.
4. Dhillon, B.S., *Engineering Maintenance: A Modern Approach*, CRC Press, Boca Raton, FL, 2002.
5. Dhillon, B.S., Rayapati, S.N., Reliability and availability analysis of transit systems, microelectronics and reliability, Vol. 25, No. 6, 1985, pp. 1073–1085.

6. Dhillon, B.S., RAM analysis of vehicles in changing weather, *Proceedings of the Annual Reliability and Maintainability Symposium*, 1984, pp. 48–53.
7. Dhillon, B.S., Kirmizi, F., Probabilistic safety analysis of maintainable systems, *International Journal of Quality in Maintenance Engineering*, Vol. 9, No. 3, 2003, pp. 303–320.
8. Dhillon, B.S., *The Analysis of the Reliability of the Multi-State Device Networks*, PhD dissertation, National Library of Canada, Ottawa, ON, 1975.
9. Fabrycky, W.J., Ghare, P.M., Torgersen, P.E., *Industrial Operations Research*, Prentice-Hall, Englewood Cliffs, New Jersey, 1972.
10. Dhillon, B.S., *Reliability Engineering in Systems Design and Operation*, Van Nostrand Reinhold Company, New York, 1983.

Appendix: Bibliography: Literature on Engineering System Reliability, Safety, and Maintenance

A.1 Introduction

Over the years, a large number of publications on engineering system reliability, safety, and maintenance have appeared in the form of journal articles, conference proceedings articles, technical reports, etc. This appendix presents an extensive list of selective publications, directly or indirectly, related to engineering system reliability, safety, and maintenance.

The period covered by the listing is from 1926 to 2013. The main objective of this listing is to provide readers with sources for obtaining additional information on engineering system reliability, safety, and maintenance.

A.2 Publications

Abbott, D.J., Application of data recording to the maintenance of railway equipment and to the improvement in reliability, *Proceedings of the International Conference on Rail Transport Systems*, 1994, pp. 123–131.

Abdi, H., Maciejewski, A.A., Nahavandi, S., Reliability maps for probabilistic guarantees of task motion for robotic manipulators, *Advanced Robotics*, Vol. 27, No. 2, 2013, pp. 81–92.

Abdul, S., Liu, G., Decentralised fault tolerance and fault detection of modular and reconfigurable robots with torque sensing, *Proceedings of the IEEE International Conference on Robotics and Automation*, 2008, pp. 3520–3526.

Abdul, S., Liu, G., Fault tolerant control of modular and reconfigurable robot with joint torque sensing, *Proceedings of the IEEE International Conference on Robotics and Biomimetics*, 2007, pp. 1236–1241.

Adams, S.K., Sabri, Z.A., Husseiny, A.A., Maintenance and testing errors in nuclear power plants: A preliminary assessment, *Proceedings of the Human Factors 24th Annual Meeting*, 1980, pp. 280–284.

Aghazadeh, F., Hirschfeld, R., Chapleski, R., Industrial robot use: Survey results and hazard analysis, *Proceedings of the 37th Annual Meeting of the Human Factors and Ergonomics Society*, 1993, pp. 994–998.

Akeel, H.A., Hardware for robotics safety systems, *Proceedings of the Inter robot West 2nd Annual Conference,* 1984, pp. 8–10.

Akeel, H.A., Intrinsic robot safety, *Proceedings of the Robot Safety Conference,* 1983, pp. 41–52.

Alayan, H., Niznik, C.A., Newcomb, R.W., Reliability of basic robot automated manufacturing networks, *Proceedings of the IEEE Southeastcon Annual Conference,* 1984, pp. 291–294.

Allen, J.O. Rankin, W.L., Use of the maintenance error decision aid (MEDA) to enhance safety and reliability and reduce costs in the commercial aviation industry, *Proceedings of the Tenth Federal Aviation Administration Meeting on Human Factors Issues in Aircraft Maintenance and Inspection: Maintenance Performance Enhancement and Technician Resource Management,* 1996, pp. 79–87.

Allen, J.P., Marx, D.M., Maintenance error decision aid project, *Proceedings of the Eighth Federal Aviation Administration Meeting on Human Factors Issues in Aircraft Maintenance and Inspection: Trends and Advances in Aviation Maintenance Operations,* 1994, pp. 101–116.

Allen, J.P., Rankin, W.L., Summary of the use and impact of the maintenance error decision aid (MEDA) on the commercial aviation industry, *Proceedings of the International Air Safety Seminar,* 1995, pp. 359–369.

Altamuro, V.M., How to achieve employee support, safety and success in your first robot installation, *Proceedings of the Robots 8 Conference,* 1984, pp. 15-1–15-8.

Altamuro, V.M., Working safely with iron collar worker, *National Safety News,* 1983, pp. 38–40.

Aluwalia, R.S., Hsu, E.Y., Sensor-based obstruction avoidance technique for a mobile robot, *Journal of Robotic Systems,* Vol. 1, 1984, pp. 331–350.

Amalberti, R., Wioland, L., Human error in aviation, *Proceeding of the International Aviation Safety Conference on Aviation Safety: Human Factors, System Engineering, Flight Operations, Economics, and Strategies Management,* 1997, pp. 91–108.

Ambs, J.L., Setren, R.S., Safety evaluation of disposable diesel exhaust filters for permissible mining equipment, *Proceedings of the Seventh US Mine Ventilation Symposium,* 1995, pp. 105–110.

American National Standard for Industrial Robots and Robot Systems-Safety Requirements, ANSI/RIA R15.06, American Standards Institute, New York, 1986.

Amir, Y., Caudy, R., Munjal, A., Schlossnagle, T., Tutu, C., N-way fail-over infrastructure for reliable servers and routers, *Proceedings of the IEEE International Conference on Dependable Systems and Networks,* 2003, pp. 130–135.

Ammar, H.H., Cukic, B., Mili, A., Fuhrman, C., A comparative analysis of hardware and software fault tolerance: Impact on software reliability engineering, *Annuals of Software Engineering,* Vol. 10, 2000, pp. 103–150.

Amrozowlcz, M.D., Brown, A., Golay, M., Probabilistic analysis of tanker groundings, *Proceedings of the International Offshore and Polar Engineering Conference,* Vol. 4, 1997, pp. 313–320.

Ancusa, V., Message redundancy in sensor networks implemented with intelligent agents, *Proceedings of the IEEE International Workshop on Robotic and Sensors Environments,* 2008, pp. 87–91.

Anderson, D.E., Malone, T.B., Baker, C.C., Recapitalizing the navy through optimized manning and improved reliability, *Naval Engineers Journal,* Vol. 110, No. 6, 1998, pp. 61–72.

Anderson, J.E., Macri, F.J., Multiple redundancy applications in a computer, *Proceedings of the Annual Symposium on Reliability*, 1967, pp. 553–562.

Anon., Automatic robot safety shutdown system, *NASA Technical Briefs*, 1984, pp. 546–547.

Anon., Cartridge valve/brake systems increases robot safety, *Robotics World*, Vol. 1, No. 4, 1983, pp. 26–28.

Anon., Computing and control division colloquium on safety and reliability of complex robotic systems, *IEE Colloquium (Digest)*, No. 085, London, 1994.

Anon., Electrostatic liquid cleaning prevents servo-valve failures, *Robotics World*, Vol. 5, 1987, pp. 32–34.

Anon., Protecting workers from robots, *American Machinist*, Vol. 128, No. 3, 1984, pp. 85–86.

Anon., Rail rudder propeller variant offers more safety in coastal shipping, *HSB International*, Vol. 41, No. 11, 1993, pp. 40–41.

Anon., Robot safety: In a state of flux and jungle, *Robot News International*, 1982, pp. 3–4.

Anon., Safety assessment for ships manoeuvring in ports, *Dock & Harbour Authority*, Vol. 79, No. 889–892, 1998, pp. 26–29.

Anon., Safety with sophistication, *The Industrial Robot*, Vol. 11, No. 4, 1984, pp. 243–245.

Anon., Sensors for robot safety, *Robotics World*, Vol. 1, No. 5, 1983, pp. 16–19.

Argote, L., Goodman, P.S., Schkade, D., The human side of robotics: How workers react to a robot, *Sloan Management Review*, 1983, pp. 33–41.

Asikin, D., Dolan, J., Reliability impact on planetary robotic missions, *Proceedings of the IEEE International Conference on Intelligent Robots and Systems*, 2010, pp. 4095–4100.

Atcitty, C.B., Robinson, D.G., Safety assessment of a robotic system handling nuclear material, *Proceedings of the ASCE Specialty Conference*, 1996, pp. 255–261.

Ayari, N., Barbaron, D., Lefevre, L., On improving the reliability of internet services through active replication, *Proceedings of the International Conference on Parallel and Distributed Computing*, 2008, pp. 259–262.

Ayrulu, B., Barshan, B., Reliability measure assignment to sonar for robust target differentiation, *Pattern Recognition*, Vol. 35, No. 6, 2002, pp. 1403–1419.

Ayyub, B.M., Lai, K., Uncertainty measures in reliability assessment of ship structures, *Proceedings of the 14th Structures Congress*, 1996, pp. 621–626.

Azadeh, A., Hasani, F.A., Jiryaei, S.Z., Performance assessment and optimization of HSE management systems with human error and ambiguity by an integrated fuzzy multivariate approach in a large conventional power plant manufacture, *Journal of Loss Prevention in the Process Industries*, Vol. 25, No. 3, 2012, pp. 594–603.

Baarman, L., Assessment of ship machinery reliability performance with numerical simulation, *Proceedings of the Conference on Numerical Tools for Ship Design*, 1995, pp. 199–212.

Bacchi, M., Cacciabue, C., O' Connor, S., Reactive and proactive methods for human factors studies in aviation maintenance, *Proceedings of the Ninth International Symposium on Aviation Psychology*, 1997, pp. 991–996.

Balkey, J.P., Human factors engineering in risk-based inspections, *Safety Engineering and Risk Analysis*, Vol. 6, 1996, pp. 97–106.

Bandyopadhyay, D., Safety management ships, *Journal of the Institution of Engineers (India), Part MR: Marine Engineering Division*, Vol. 84, No. 2, 2003, pp. 45–48.

Banker, R.D., Datar, S.M., Kemerer, C.F., Software errors and software maintenance management, *Information Technology and Management*, Vol. 3, No. 1–2, 2002, 25–41.

Barabanov, V., Chirkov, A.L., Reliability of robot assisted manufacturing cells in their initial operating period, *Soviet Engineering Research*, Vol. 10, No. 1, 1990, pp. 83–85.

Barcheck, C., Methods for safe robotics start-up, testing, inspection and maintenance, *Proceedings of the RIA Robot Seminar*, 1985, pp. 67–73.

Barczak, T., Gearhart, D., Performance and safety considerations of hydraulic support systems, *Proceedings of the 17th International Conference on Ground Control in Mining*, 1998, pp. 176–186.

Barrett, R.J., Bell, R., Hodson, P.H., Planning for robot installation and maintenance: A safety framework, *Proceedings of the 4th British Robot Association Annual Conference*, 1981, pp. 13–22.

Barrett, R.J., Robot safety and the law, in *Robot Safety*, edited by M.C. Bonney, Y.F. Yong, Springer-Verlag, Berlin, 1985, pp. 3–15.

Bayley, J.M., Uber, C.B., Comprehensive program to improve road safety at railway level crossings, *Proceedings of the 15th ARRB Conference*, 1990, pp. 217–234.

Benham, H., Spreadsheet structure and maintenance errors, *Proceedings of the Information Resources Management Association International Conference*, 1993, pp. 262–267.

Berg, H., Wolfgang, E., Advanced IGBT modules for railway traction applications: Reliability testing, *Microelectronics and Reliability*, Vol. 38, No. 6–8, 1998, pp. 1319–1323.

Blache, K.M., Industrial practices for robotic safety, in *Safety, Reliability, and Human Factors*, edited by J.H. Graham, Van Nostrand Reinhold, New York, 1991, pp. 34–65.

Boland, P.J., Singh, H., A birth-process approach to Moranda's geometric software reliability model, *IEEE Transactions on Reliability*, Vol. 52, 2003, pp. 168–174.

Boring, R.L., Modeling human reliability analysis using MIDAS, *Proceedings of the International Workshop on Future Control Station Designs and Human Performance Issues in Nuclear Power Plants*, 2009, pp. 89–92.

Borse, E., Design basis accidents and accident analysis with particular reference to offshore platforms, *Journal of Occupational Accidents*, Vol. 2, No. 3, 1979, pp. 227–243.

Bouillant, L., Weber, P., Salem, A.B., Aknin, P., Use of causal probabilistic networks for the improvement of the maintenance of railway infrastructure, *Proceedings of the International Conference on Systems, Man, and Cybernetics*, 2004, pp. 6243–6249.

Bousvaros, G.A., Don, C., Hopps, J.A., An electrical hazard of selective angiocardiography, *Canadian Medical Association Journal*, Vol. 87, 1962, pp. 286–288.

Braithwaite, G.R., A safe culture or safety culture, *Proceedings of the Ninth International Symposium on Aviation Psychology*, 1997, pp. 1029–1031.

Brewer, B.R., Pradhan, S., Preliminary investigation of test-retest reliability of a robotic assessment for Parkinson's disease, *Proceedings of the 32nd Annual International Conference of the IEEE EMBS*, 2010, pp. 5863–5866.

Brown, W.R., Ulsoy, A.G., A passive-assist design approach for improved reliability and efficiency of robots arms, *Proceedings of the IEEE International Conference on Robotics and Automation*, 2011, pp. 4927–4934.

Burchell, H.B., Electrocution hazards in the hospital or laboratory, *Circulation*, 1963, pp. 1015–1017.

Butani, S. J., Hazard analysis of mining equipment by mine type and geographical region, *Proceedings of the Symposium on Engineering Health and Safety*, 1986, pp. 158–173.

Calestine, A., Park, E.H., A formal method to characterize robot reliability, *Proceedings of the Annual Reliability and Maintainability Symposium*, 1993, pp. 395–398.

Cangussu, J.W., Mathur, A.P., Karcich, R.M., Software release control using defect based quality estimation, *Proceedings of the 15th International Symposium on Software Reliability Engineering*, 2004, pp. 440–450.

Card, D.N., Managing software quality with defects, *Proceedings of the 26th Annual International Computer Software and Applications Conference*, 2002, pp. 472–474.

Carlson, J., Murphy, R.R., Reliability analysis of mobile robots, *Proceedings of the IEEE International Conference on Robotics and Automation*, 2003, pp. 247–281.

Carlson, R., Hobby, R., Newman, H.B., measuring end-to-end internet performance, *Network Magazine*, Vol. 18, No. 4, 2003, pp. 42–46.

Caro, A. et al., A proposal for a set of attributes relevant for web portal data quality, *Software Quality Journal*, Vol. 16, No. 4, 2008, pp. 513–542.

Carr, M.J., Christer, A.H., Incorporating the potential for human error in maintenance models, *Journal of the Operational Research Society*, Vol. 54, No. 12, 2003, pp. 1249–1253.

Carreras, C., Walker, I.D., An interval method applied to robot reliability quantification, *Reliability Engineering and System Safety*, Vol. 70, 2000, pp. 291–303.

Carreras, C., Walker, I.D., Interval methods for fault-tree analysis in robotics, *IEEE Transactions on Reliability*, Vol. 50, No. 1, 2001, pp. 3–11.

Carreras, C., Walker, I.D., Interval methods for improved robot reliability estimation, *Proceedings of the Annual Reliability and Maintainability Symposium*, 2000, pp. 22–27.

Cavalier, M.P., Knapp, G.M., Reducing preventive maintenance cost error caused by uncertainty, *Journal of Quality in Maintenance Engineering*, Vol. 2, No. 3, 1996, pp. 21–36.

Cawley, J.C., Electrical accidents in the mining industry, 1990–1999, *IEEE Transactions on Industry Applications*, Vol. 39, No. 6, 2003, pp. 1570–1577.

Celeux, G., Corset, F., Garnero, M.A., Breuils, C., Accounting for inspection errors and change in maintenance behaviour, *IMA Journal of Management Mathematics*, Vol. 13, No. 1, 2002, pp. 51–59.

Cha, S.S., Management aspect of software safety, *Proceedings of the Eighth Annual Conference on Computer Assurance*, 1993, pp. 35–40.

Chen, S.K., Ho, T.K., Mao, B.H., Reliability evaluations of railway power supplies by fault tree analysis, *IET Electric Power Applications*, Vol. 1, No. 2, 2007, pp. 161–172.

Cheng-Yi, L., Optimization of maintenance, production and inspection strategies while considering preventive maintenance error, *Journal of Information and Optimization Sciences*, Vol. 25, No. 3, 2004, pp. 543–555.

Chien-Kuo, S., Cheng-Yi, L., Optimizing an integrated production and quality strategy considering inspection and preventive maintenance errors, *Journal of Information and Optimization Sciences*, Vol. 27, No. 3, 2006, pp. 577–593.

Christer, A.H., Lee, S.K., Modelling ship operational reliability over a mission under regular inspections, *Journal of the Operational Research Society*, Vol. 48, No. 7, 1997, pp. 688–699.

Chung, W.K., Reliability analysis of a repairable parallel system with standby involving human error and common-cause failure, *Microelectronics and Reliability*, Vol. 27, No. 2, 1987, pp. 269–271.

Crowder, R.M. et al., Maintenance of robotic systems using hypermedia and case-based reasoning, *Proceedings of the IEEE International Conference on Robotics and Automation*, 2000, pp. 2422–2427.

Danahar, J.W., Maintenance and inspection issues in aircraft 87 accidents/incidents, *Proceedings of the Meeting on Human Factors Issues in Aircraft Maintenance and Inspection*, 1989, pp. A9–A11.

Daniel, J.H., Reducing mine accidents by design, *Proceedings of the SME Annual Meeting*, 1991, pp. 1–11.

Darwish, M.A., Ben-Daya, M., Effect of inspection errors and preventive maintenance on a two-stage production inventory system, *International Journal of Production Economics*, Vol. 107, No. 1, 2007, pp. 301–303.

Davies, R.K.L., Monitoring hoist safety equipment, *Coal Age*, Vol. 21, No. 1, 1986, pp. 66–68.

Denzler, H.E., How safety is designed into offshore platforms, *World Oil*, Vol. 152, No. 7, 1961, pp. 131–133, 136.

Desai, M. et al., Effects of changing reliability on trust of robot systems, *Proceedings of the IEEE International Conference on Human-Robot Interaction*, 2012, pp. 73–80.

Dhananjay, K., Bengt, K., Uday, K., Reliability analysis of power transmission cables of electric mine loaders using the proportional hazards model, *Reliability Engineering & System Safety*, Vol. 37, No. 3, 1992, pp. 217–222.

Dhillon, B.S., Aleem, M.A., Report on robot reliability and safety in canada: A survey of robot users, *Journal of Quality in Maintenance Engineering*, Vol. 6, No. 1, 2000, pp. 61–74.

Dhillon, B.S., Anude, O.C., Robot safety and reliability: A review, *Microelectronics and Reliability*, Vol. 33, No. 3, 1993, pp. 413–429.

Dhillon, B.S., *Design Reliability*, CRC Press, Boca Raton, FL, 1999.

Dhillon, B.S., *Engineering Maintenance: A Modern Approach*, CRC Press, Boca Raton, FL, 2002.

Dhillon, B.S., Fashandi, A.R.M., Liu, K.L., Robot systems reliability and safety: A review, *Journal of Quality in Maintenance Engineering*, Vol. 8, No. 3, 2002, 170–212.

Dhillon, B.S., Fashandi, A.R.M., Safety and reliability assessment techniques in robotics, *Robotica*, Vol. 15, No. 6, 1997, pp. 701–708.

Dhillon, B.S., Fashandi, A.R.M., Stochastic analysis of a robot machine with duplicate safety units, *Journal of Quality in Maintenance Engineering*, Vol. 5, No. 2, 1999, pp. 114–127.

Dhillon, B.S., Human error in medical systems, *Proceedings of the 6th ISSAT International Conference on Reliability and Quality in Design*, 2000, pp. 138–143.

Dhillon, B.S., Li, Z., Stochastic analysis of a maintainable robot-safety system with common-cause failures, *Journal of Quality in Maintenance Engineering*, Vol. 10, No. 2, 2004, pp. 136–147.

Dhillon, B.S., Liu, Y., Human error in maintenance: A review, *Journal of Quality in Maintenance Engineering*, Vol. 12, No. 1, 2006, pp. 21–36.

Dhillon, B.S., *Medical Device Reliability and Associated Areas*, CRC Press, Boca Raton, FL, 2000.

Dhillon, B.S., Modeling human errors in repairable systems, *Proceedings of the Annual Reliability and Maintainability Symposium*, 1989, pp. 418–424.

Dhillon, B.S., On robot reliability and safety: Bibliography, *Microelectronics and Reliability*, Vol. 27, 1987, pp. 105–118.

Dhillon, B.S., *Robot Reliability and Safety*, Springer-Verlag, New York, 1991.

Dhillon, B.S., Yang, N., Probabilistic analysis of a maintainable system with human error, *Journal of Quality in Maintenance Engineering*, Vol. 1, No. 2, 1995, pp. 50–59.

Dhillon, B.S., Yang, N.F., Human error analysis of a standby redundant system with arbitrarily distributed repair times, *Microelectronics and Reliability*, Vol. 33, No. 3, 1993, pp. 431–444.

Douglass, D.P., Safety devices on electric hoists used in Ontario mines, *Canadian Mining Journal*, Vol. 61, No. 4, 1940, pp. 229–234.

Drury, C.G., Shepherd, W.T., Johnson, W.B., Error reduction in aviation maintenance, *Proceedings of the 13th Triennial Congress of the International Ergonomics Association*, 1997, pp. 31–33.

Duffner, D.H., Torsion fatigue failure of bus drive shafts, *Journal of Failure Analysis and Prevention*, Vol. 6, No. 6, 2006, pp. 75–82.

DuPont, G., The dirty dozen errors in maintenance, *Proceedings of the 11th FAA/AAM Meeting on Human Factors in Aviation Maintenance and Inspection*, 1997, pp. 49–52.

El Koursi, E., Flahaut, G., Zaalberg, H., Hessami, A., Safety assessment of European rail rules for operating ERTMS, *Proceedings of the International Conference on Automated People Movers*, 2001, pp. 811–815.

El-Badan, A., Leheta, H.W., Abdel-Nasser, Y., Moussa, M.A., Safety assessment for ship hull girders taking account of corrosion effects, *Alexandria Engineering Journal*, Vol. 41, No. 1, 2002, pp. 71–81.

Etherton, J.R., Systems considerations on robot and effector speed as a risk factor during robot maintenance, *Proceedings of the 8th International Safety Conference*, 1987, pp. 343–347.

Fink, R.A., Reliability modeling of freely-available internet-distributed software, *Proceedings of the International Software Metrics Symposium*, 1998, pp. 101–104.

Fischer, A.L., Camera sensors boost safety for Japanese railway, *Photonics* Spectra, Vol. 38, No. 9, 2004, pp. 33–34.

Ford, T., Three aspects of aerospace safety-human factors in airline maintenance, *Aircraft Engineering and Aerospace Technology*, Vol. 69, No. 3, 1997, pp. 262–264.

Fortier, S.C., Michael, J.B., A risk-based approach to cost–benefit analysis of software safety activities, *Proceedings of the Eighth Annual Conference on Computer Assurance*, 1993, pp. 53–60.

Fox, D., Robotic safety, *Robotics World*, January/February 1999, pp. 26–29.

Freihrr, G., Safety is key to product quality, productivity, *Medical Device & Diagnostic Industry Magazine*, Vol. 19, No. 4, 1997, pp. 18–19.

Fries, R.C., Pienkowski, P., Jorgens, J., Safe, effective and reliable software design and development for medical devices, *Medical Instrumentation*, Vol. 30, No. 2 1996, pp. 75–80.

Fuller, D.A., Managing risk in space operations: Creating and maintaining a high reliability organization, *Proceedings of the AIAA Space Conference*, 2004, pp. 218–223.

Genova, R., Galaverna, M., Sciutto, G., Zavatoni, V., Techniques for human performance analysis in railway applications, *Proceedings of the International Conference on Computer Aided Design, Manufacture and Operation in the Railway and Other Advanced Mass Transit Systems*, 1998, pp. 959–968.

Gerdun, V., Sedmak, T., Sinkovec, V., Kovse, I., Cene, B., Failures of bearings and axles in railway freight wagons, *Engineering Failure Analysis*, Vol. 14, No. 5, 2007, pp. 884–894.

Gibson, C.S., Aspects of safety affecting mechanical and electrical services in mines, *Canadian Mining Journal*, Vol. 73, No. 5, 1952, pp. 66–71.

Goseva-Popstojanova, K., Mazimdar, S., Singh, A.D., Empirical study of session-based workload and reliability for web servers, *Proceedings of the International Symposium on Software Reliability Engineering*, 2004, pp. 403–414.

Goseva-Popstojanova, K., Trivedi, K.S., Architecture-based approaches to software reliability prediction, *Computers and Mathematics with Applications*, Vol. 46, No. 7, 2003, pp. 1023–1036.

Gowen, L.D., Yap, M.Y., Traditional software development's effects on safety, *Proceedings of the 6th Annual IEEE Symposium on Computer-Based Medical Systems*, 1993, pp. 58–63.

Graeber, R.C., Marx, D.A., Reducing human error in aircraft maintenance operations, *Proceeding of the 46th Annual International Air Safety Seminar & International Federation of Airworthiness 23rd International Conference*, November 8–11, 1993, pp. 147–157.

Graham, J.H., Overview of robot safety, reliability, and human factors issues, in *Safety, Reliability, and Human Factors in Robotic Systems*, edited by J.H. Graham, Van Nostrand Reinhold, 1991, pp. 1–10.

Gramopadhye, A.K., Drury, C.G., Human factors in aviation maintenance: How we get to where we are? *International Journal of Industrial Ergonomics*, Vol. 26, No. 2, 2000, pp. 125–131.

Guedes, S.C., Garbatov, Y., Reliability of maintained ship hulls subjected to corrosion, *Journal of Ship Research*, Vol. 40, No. 3, 1996, pp. 495–516.

Hadianfard, M.J., Hadianfard, M.A., Structural failure of a telescopic shiploader during installation, *Journal of Failure Analysis and Prevention*, Vol. 7, No. 4, 2007, pp. 282–291.

Han, L.D., Simulating ITS operations safety with virtual reality, *Proceedings of the Transportation Congress*, Vol. 1, 1995, pp. 215–226.

Hansen, M., Zhang, Y., Safety of efficiency: Link between operational performance and operational errors in the national airspace system, *Transportation Research Record*, No. 1888, 2004, pp. 15–21.

Hecht, M., Reliability/availability modeling, prediction, and measurement for e-commerce and other internet information systems, *Proceedings of the Annual Reliability and Maintainability Symposium*, 2001, pp. 176–182.

Hee, D.D., Pickrell, B.D., Bea, R.G., Roberts, K.H., Williamson, R.B., Safety management assessment system (SMAS): A process for identifying and evaluating human and organization factors in marine system operations with field test results, *Reliability Engineering and System Safety*, Vol. 65, No. 2, 1999, pp. 125–140.

Heinrich, D.J., Safer approaches and landings: A multivariate analysis of critical factors, *Proceedings of the Corporate Aviation Safety Seminar*, 2005, pp. 103–155.

Heyns, F., Van Der Westhuizen, J., A mining case study: The safe maintenance of underground railway track, *Civil Engineering*, Vol. 14, No. 5, 2006, pp. 8–10.

Heyns, F.J., Construction and maintenance of underground railway tracks to safety standard of SANS: 0339, *Journal of the South African Institute of Mining and Metallurgy*, Vol. 106, No. 12, 2006, pp. 793–798.

Hibit, R., Marx, D.A., Reducing human error in aircraft maintenance operations with the maintenance error decision aid (MEDA), *Proceedings of the Human Factors and Ergonomics Society 38th Annual Meeting*, 1994, pp. 111–114.

Hidaka, H., Yamagata, T., Suzuki, Y., Structuring a new maintenance system, *Japanese Railway Engineering*, No. 132–133, 1995, pp. 7–10.

Hirschfeld, R.A., Aghazadeh, F., Chapleski, R.C., Survey of robot safety in industry, *International Journal of Human Factors in Manufacturing*, Vol. 3, No. 4, 1993, pp. 369–379.

Hobbs, A., Maintenance mistakes and system solutions, *Asia Pacific Air Safety*, Vol. 21, 1999, pp. 1–7.

Hopps, J.A., Electrical hazards in hospital instrumentation, *Proceedings of the Annual Symposium on Reliability*, 1969, pp. 303–307.

Howard, S., Hammond, J., Lindgaard, G., editors, *Human–Computer Interaction: Interact'97*, Chapman and Hall, London, 1997.

Huang, H., Yuan, X., Yao, X., Fuzzy fault tree analysis of railway traffic safety, *Proceedings of the Conference on Traffic and Transportation Studies*, 2000, pp. 107–112.

Huang, W.G., Zhang, L., Cause analysis and preventives for human error events in Daya Nay NPP, *Dongli Gongcheng/Nuclear Power Engineering*, Vol. 19, No. 1, 1998, pp. 64–67, 76.

Hudoklin, A., Rozman, V., Human errors versus stress, *Reliability Engineering & System Safety*, Vol. 37, 1992, pp. 231–236.

Hudoklin, A., Rozman, V., Reliability of railway traffic personnel, *Reliability Engineering & System Safety*, Vol. 52, 1996, pp. 165–169.

Hudoklin, A., Rozman, V., Safety analysis of the railway traffic system, *Reliability Engineering & System Safety*, Vol. 37, 1992, pp. 7–13.

Hughes, S., Warner Jones, S., Shaw, K., Experience in the analysis of accidents and incidents involving the transport of radioactive materials, *Nuclear Engineer*, Vol. 44, No. 4, 2003, pp. 105–109.

Husband, T.M., Managing robot maintenance, *Proceedings of the 6th British Robot Association Annual Conference*, 1983, pp. 53–60.

Hyman, W.A., Errors in the use of medical equipment, In *Human Error in Medicine*, edited by M.S. Bogner, L. Erlbaum Associates Publishers, Hillsdale, NJ, 1994, pp. 327–348.

IEC 601-1: Safety of Medical Electrical Equipment, Part 1: General Requirements, International Electrotechnical Commission (IEC), Geneva, 1977.

IEEE-STD-1228, Standard for Software Safety Plans, Institute of Electrical and Electronic Engineers (IEEE), New York, 1994.

Ikuta, K., Nokata, M., General evaluation method of safety for human-care robots, *Proceedings of the IEEE International Conference on Robotics and Automation*, 1999, pp. 2065–2072.

Inagaki, S., Sato, K., Summary report on the status of safety engineering for industrial robots in the united states, *Katakana/Robot*, Vol. 114, 1997, pp. 17–21.

Industrial robots and robot system safety, in *OSHA Technical Manual*, Occupational Safety and Health Administration, Department of Labor, Washington, DC, 2001.

Ingleby, M., Mitchell, I., Proving safety of a railway signalling system incorporating geographic data, *Proceedings of the IFAC Symposium*, 1992, pp. 129–134.

Inoue, T., Kusukami, K., Kon-No, S., Car driver behavior in railway crossing accident, *Quarterly Report of RTRI (Railway Technical Research Institute of Japan)*, Vol. 37, No. 1, 1996, pp. 26–31.

Inozu, B., Schaedel, P.G., Molinari, V., Roy, P., Johns, R., Reliability data collection for ship machinery, *Marine Technology*, Vol. 35, No. 2, 1998, pp. 119–125.

Ippolito, L.M., Wallace, D.R., *A Study on Hazard Analysis in High Integrity Software Standards and Guidelines*, Report No. NISTIR 5589, National Institute of Standards and Technology, US Department of Commerce, Washington, DC, January 1995.

Iskander, W.H., Nutter, R.S., Methodology development for safety and reliability analysis for electrical mine monitoring systems, *Microelectronics and Reliability*, Vol. 28, No. 4, 1988, pp. 581–597.

Isoda, H., Yasutake, J.Y., Human factors interventions to reduce human errors and improve productivity in maintenance tasks, *Proceedings of the International Conference on Design and Safety of Advanced Nuclear Power Plants*, 1992, pp. 34.4/1–6.

Isozaki, Y., Ohnishi, K., Nagashima, T., New generation maintenance and inspection robots for nuclear power plant, *Proceedings of the International Conference on Offshore Mechanics and Arctic Engineering*, 1997, pp. 77–84.

Jacobsen, T., A potential of reducing the risk of ship casualties by 50%, *Marine and Maritime*, Vol. 3, 2003, pp. 171–181.

Jacobsson, L., Svensson, O., Psychosocial work strain of maintenance personnel during annual outage and normal operation in a nuclear power plant, *Proceedings of the Human Factors Society 35th Annual Meeting*, Vol. 2, 1991, pp. 913–917.

Janota, A., Using Z specification for railway interlocking safety, *Periodica Polytechnica Transportation Engineering*, Vol. 28, No. 1–2, 2000, pp. 39–53.

Jauw, J., Vassiliou, P., Field data is reliability information: Implementing an automated data acquisition and analysis system, *Proceedings of the Annual Reliability and Maintainability Symposium*, 2000, pp. 86–93.

Ji, Q., Zhu, Z., Lan, P., Real-time nonintrusive monitoring and prediction of driver fatigue, *IEEE Transactions on Vehicular Technology*, Vol. 53, No. 4, 2004, pp. 1052–1068.

Jiang, B.C., Gainer, C.A., A cause and effect analysis of robot accidents, *Journal of Occupational Accidents*, Vol. 9, 1987, pp. 27–45.

Joel, K., Duncan, S., A practical approach to fire hazard analysis for offshore structures, *Journal of Hazardous Materials*, Vol. 104, No. 1–3, 2003, pp. 107–122.

Johnson, W.B., Rouse, W.B., Analysis and classification of human error in troubleshooting live aircraft power plants, *IEEE Transactions on Systems, Man, and Cybernetics*, Vol. 12, No. 3, 1982, pp. 389–393.

Johnson, W.B., Human factors in maintenance: An emerging system requirement, *Ground Effects*, Vol. 2, 1997, pp. 6–8.

Joshi, V.V., Kaufman, L.M., Giras, T.C., Human behavior modeling in train control systems, *Proceedings of the Annual Reliability and Maintainability Symposium*, 2001, pp. 183–188.

Kamiyama, M., Furukawa, A., Yoshimura, A., The effect of shifting errors when correcting track irregularities with a heavy tamping machine, *Advances in Transport*, Vol. 7, 2000, pp. 95–104.

Kanki, B., Managing procedural error in maintenance, *Proceedings of the Flight Safety Foundation Annual International Air Safety Seminar*, 2005, pp. 233–244.

Kantowitz, B.H., Hanowski, R.J., Kantowitz, S.C., Driver acceptance of unreliable traffic information in familiar and unfamiliar settings, *Human Factors*, Vol. 39, No. 2, 1997, pp. 164–174.

Karwowski, W., Parsei, H.R., Amarnath, B., Rahimi, M., A study of worker intrusion in robots work envelope, in *Safety, Reliability, and Human Factors in Robotic Systems*, edited by J.H. Graham, Van Nostrand Reinhold, New York, 1991, pp. 148–162.

Kecojevic, V., Radomsky, M., The causes and control of loader and truck related fatalities in surface mining operations, *Injury Control and Safety Promotion*, Vol. 11, No. 1, 2004, pp. 239–251.

Keene, S., Lane, C., Combined hardware and software aspects of reliability, *Quality and Reliability Engineering International*, Vol. 8, No. 5, 1992, pp. 419–426.

Keene, S.J., Assuring software safety, *Proceedings of the Annual Reliability and Maintainability Symposium*, 1992, pp. 274–279.

Keran, C.M., Hendricks, P.A., Automation & safety of mobile mining equipment, *Engineering and Mining Journal*, Vol. 196, No. 2, 1995, pp. 30–33.

Kerstholt, J.H., Passenier, P.O., Houttuin, K., Schuffel, H., Effect of a priori probability and complexity on decision making in a supervisory control task, *Human Factors*, Vol. 38, No. 1, 1996, pp. 65–79.

Khan, F.I., Haddara, M.R., Risk-based maintenance of ethylene oxide production facilities, *Journal of Hazardous Materials*, Vol. 108, No. 3, 2004, pp. 147–159.

Kirby, M.J., Klein, R.L., Separation of maintenance and operator errors from equipment failures, *Proceedings of the Product Assurance Conference and Technical Exhibit*, 1969, pp. 17–27.

Kirwan, B., The role of the controller in the accelerating industry of air traffic management, *Safety Science*, Vol. 37, No. 2–3, 2001, pp. 151–185.

Klein, R., The human factors impact of an export system based reliability centered maintenance program, *Proceedings of the IEEE Conference on Human Factors and Power Plants*, 1992, pp. 241–245.

Knee, H.E., The maintenance personnel performance simulation (MAPPS) model: A human reliability analysis tool, proceedings of the international conference on nuclear power plant aging, *Availability Factor and Reliability Analysis*, 1985, pp. 77–80.

Kobylinski, L.K., Rational approach to ship safety requirements, *Proceedings of the International Conference on Marine Technology*, 1997, pp. 3–13.

Komarniski, R., Maintenance human factors and the organization, *Proceedings of the Corporate Aviation Safety Seminar*, 1999, pp. 265–267.

Kovari, B., Air crew training, human factors and reorganizing in case of irregularities, *Periodica Polytechnica Transportation Engineering*, Vol. 33, No. 1–2, 2005, pp. 77–88.

Kraft, E.R., A hump sequencing algorithm for real time management of train connection reliability, *Journal of the Transportation Research Forum*, Vol. 39, No. 4, 2000, pp. 95–115.

Kuenzi, J.K., Nelson, B.C., Mobile mine equipment maintenance safety: A review of U.S. Bureau of mines research, Bureau of Mines, United States Department of the Interior, Washington, DC, 1995.

Kwitowski, A.J., Brautigam, A.L., Monaghan, W.D., Teleoperated continuous mining machine for improved safety, *Mining Engineering*, Vol. 47, No. 8, 1995, pp. 753–759.

Lamonde, F., Safety Improvement in Railways: Which criteria for coordination at a distance design? *International Journal of Industrial Ergonomics*, Vol. 17, No. 6, 1996, pp. 481–497.

Larue, C., Cohen, H.H., Consumer perception of light truck safety, *Proceedings of the Human Factors Society 34th Annual Meeting*, 1990, pp. 589–590.

Latorella, K.A., Prabhu, P.V., Review of human error in aviation maintenance and inspection, *International Journal of Industrial Ergonomics*, Vol. 26, No. 2, 2000, pp. 133–161.

Layton, C.F., Shepherd, W.T., Johnson, W.B., Norton, J.E., Enhancing human reliability with integrated information systems for aviation maintenance, *Proceedings of the Annual Reliability and Maintainability Symposium*, 1993, pp. 498–502.

Lee, S., Kim, J., Choi, J., Development of a web-based integrity evaluation system for primary components in a nuclear power plant, *Proceedings of the Asian Pacific Conference on Nondestructive Testing*, 2004, pp. 2226–2231.

Leffingwell, D.A., Norman, B., Software Quality in Medical Devices: A top-down approach, *Proceedings of the 6th Annual IEEE Symposium on Computer-Based Medical Systems*, 1993, pp. 307–31.

Leveson, N.G., Shimeall, T.J., Safety verification of ADA programs using software fault trees, *IEEE Software*, July 1991, pp. 48–59.

Levkoff, B., Increasing safety in medical device software, *Medical Device & Diagnostic Industry Magazine*, Vol. 18, No. 9, 1996, pp. 92–97.

Li, D., Li, L., Preliminary study of human factor reliability in hydropower station, *Advanced Materials Research*, Vol. 422, 2012, pp. 803–806.

Li, D., Tang, W., Zhang, S., Hybrid event tree analysis of ship grounding probability, *Proceedings of the International Conference on Offshore Mechanics and Arctic Engineering—OMAE*, Vol. 2, 2003, pp. 345–349.

Li, D., Tang, W., Zhang, S., Hybrid event tree calculation of ship grounding probability caused by piloting failure, *Shanghai Jiaotong Daxue Xuebao/Journal of Shanghai Jiaotong University*, Vol. 37, No. 8, 2003, pp. 1146–1150.

Lim, A., Advanced techniques for maintaining reliability of complex computer systems, *Proceedings of the Hawaii International Conference on System Sciences*, 1997, pp. 4–13.

Lin, L.J., Cohen, H.H., Accidents in the trucking industry, International *Journal of Industrial Ergonomics*, Vol. 20, 1997, pp. 287–300.

Ling, J., Integrated reliability development for the front suspension systems in light duty dodge ram truck, *International Journal of Reliability, Quality, and Safety Engineering*, Vol. 13, No. 3, 2006, pp. 193–202.

Lipowczan, A., Increasing the reliability and safety of mining machines by application of the vibration diagnostic (experiences and results), *Proceedings of the International Conference on Reliability, Production, and Control in Coal Mines*, 1991, pp. 155–163.

Lobb, B., Harre, N., Suddendorf, T., An evaluation of a suburban railway pedestrian crossing safety programme, *Accident Analysis and Prevention*, Vol. 33, No. 2, 2001, pp. 157–165.

Lourens, P.F., Theoretical perspectives on error analysis and traffic behaviour, *Ergonomics*, Vol. 33, No. 10–11, 1990, pp. 1251–1263.

Lyon, E., Miners' electric safety lamps, *Electrical Review*, Vol. 98, No. 2510, 1926, pp. 9–10.

Lyu, M.R., Editor, *Handbook of Software Reliability Engineering*, McGraw-Hill Book Company, New York, 1996.

MacGregor, C., Hopfl, H.D., Integrating safety and systems: The implications for organizational learning, *Proceedings of the International Air Safety Seminar*, 1992, pp. 304–311.

Maddox, M.E., Designing medical devices to minimize human error, *Medical Device & Diagnostic Magazine*, Vol. 19, No. 5, 1997, pp. 166–180.

Maillart, L.M., Pollock, S.M., The effect of failure-distribution specificationerrors on maintenance costs, *Proceedings of the Annual Reliability and Maintainability Symposium*, 1999, pp. 69–77.

Majos, K., Communication and operational failures in the cockpit, *Human Factors and Aerospace Safety*, Vol. 1, No. 4, 2001, p. 323–340.

Majumdar, A., Ochieng, W.Y., Nalder, P., Trend analysis of controller-caused airspace incidents in New Zealand, 1994–2002, *Transportation Research Record*, No. 1888, 2004, pp. 22–33.

Majumdar, A., Ochleng, W.Y., Nalder, P., Airspace safety in New Zealand: A causal analysis of controller caused airspace incidents between 1994–2002, *The Aeronautical Journal*, Vol. 108, May 2004, pp. 225–236.

Malone, T.B., Rousseau, G.K., Malone, J.T., Enhancement of human reliability in port and shipping operations, *Water Studies*, Vol. 9, 2000, pp. 101–111.

Marshall, K.L., Mine safety as affected by electrification, *Coal Mining*, Vol. 5, No. 3, 1928, pp. 79–80.

Marx, D.A., Moving toward 100% error reporting in maintenance, *Proceedings of the 11th FAA/AAM Meeting on Human Factors in Aviation Maintenance and Inspection*, 1997, pp. 42–48.

Mathur, F.P., On reliability modelling and analysis of a dynamic tmr system utilizing standby spares, *Proceedings of the Seventh Annual Allerton Conference on Circuits and Systems*, 1969, pp. 115–120.

McDowell, A., Schmidt, C., Yue, K., Analysis and metrics of XML schema, *Proceedings of the International Conference on Software Engineering Research and Practice*, 2004, pp. 538–544.

McGrath, R.N., Safety and maintenance management: A View from an ivory tower, *Proceedings of the Aviation Safety Conference and Exposition*, 1999, pp. 21–26.

Meadow, L., Los Angeles Metro blue line light rail safety issues, *Transportation Research Record*, No. 1433, 1994, pp. 123–133.

Menon, D., Kumaran, G., Dynamic analysis and reliability-based analysis of PSC rail-track sleepers, *Journal of Structural Engineering*, Vol. 30, No. 1, 2003, pp. 25–31.

Metzger, U., Parasuraman, R., Automation in future air traffic management: Effects of decision aid reliability on controller performance and mental workload, *Human Factors*, Vol. 47, No. 1, 2005, pp. 35–49.

Meulen, M.V.D., *Definitions for Hardware and Software Safety Engineers*, Springer-Verlag, London, 2000.

Meyer, J.L., Some instrument-induced errors in the electrocardiogram, *Journal of the American Medical Association*, Vol. 201, 1967, pp. 351–358.

Mikulski, J., Malfunctions of railway traffic control systems-failure rate analysis, *Proceedings of the 3rd International Conference on Computer Simulation in Risk Analysis and Hazard Mitigation*, 2002, p. 141–147.

Mitra, A.K., Relevance of condition based maintenance onboard a ship, *Journal of the Institution of Engineers (India), Part MR: Marine Engineering Division*, Vol. 86, 2006, pp. 50–55.

Mojdehbakhsh, R., Tsai, W.T., Kirani, S., Elliott, L., Retrofitting software safety in an implantable medical device, *IEEE Software*, No. 1, Jan. 1994, pp. 41–50.

Morgenstern, M.H., Maintenance management systems: A human factors issue, *Proceedings of the IEEE Conference on Human Factors and Power Plants*, 1988, pp. 390–393.

Morrell, H.W., European standards—Mining machinery safety, *Mining Technology*, Vol. 74, No. 851, 1992, pp. 13–14.

Murakami, M., Development of a duplex computer system for humanoid robot applications: Design of the safety failover subsystem, *Proceedings of the 33rd Annual Conference of the IEEE Industrial Electronics Society*, 2007, pp. 2783–2788.

Murakami, M., Fault tolerance design for computers used in humanoid robots, *Proceedings of the International Conference on Intelligent Robots and Systems*, 2007, pp. 2301–2307.

Nagamachi, M., Human factors in industrial robots: Robot safety management in Japan, *Applied Ergonomics*, Vol. 17, No. 1, 1986, pp. 9–18.

Nagamachi, M., Ten fatal accidents due to robots in Japan, in *Ergonomics of Hybrid Automated Systems*, edited by W. Karwowski et al., Elsevier, Amsterdam, 1988, pp. 391–396.

Neal, M.L., Managing software quality through defect trend analysis, *Proceedings of the PMI Annual Seminar/Symposium*, 1991, pp. 119–122.

Nelson, D., O'Neil, K., Commuter rail service reliability: On-time performance and causes for delays, *Transportation Research Record*, No. 1704, 2000, pp. 42–50.

Nianfu, Y., Dhillon, B.S., Stochastic analysis of a general standby system with constant human error and arbitrary system repair rates, *Microelectronics and Reliability*, Vol. 35, No. 7, 1995, pp. 1037–1045.

Nibbering, J.J.W., Structural safety and fatigue of ships, *International Shipbuilding Progress*, Vol. 39, No. 420, 1992, pp. 61–98.

Nicolaisen, P., Ways of improving industrial safety for the programming of industrial robots, *Proceedings of the 3rd International Conference on Human Factors in Manufacturing*, November 1986, pp. 263–276.

Niu, X., Huang, X., Zhao, Z., Zhang, Y., Huang, C., Cui, L., The design and evaluation of a wireless sensor network for mine safety monitoring, *Proceedings of the IEEE Global Telecommunications Conference*, 2007, pp. 1291–1295.

Nobel, J.L., Medical device failures and adverse effects, *Pediatric Emergency Care*, Vol. 7, 1991, pp. 120–123.

Novak, M., Votruba, Z., Challenge of human factor influence for car safety, *Neural Network World*, Vol. 14, No. 1, 2004, pp. 37–41.

Novak, M., Votruba, Z., Faber, J., Impacts of driver attention failures on transport reliability and safety and possibilities of its minimizing, *Neural Network World*, Vol. 14, No. 1, 2004, pp. 49–65.

Nunn, R., Witts, S.A., The influence of human factors on the safety of aircraft maintenance, *Proceedings of the Flight Safety Foundation/International Federation of Airworthiness/Aviation Safety Conference*, 1997, pp. 212–221.

O'Leary, M., Chappell, S., Confidential incidents reporting systems create vital awareness of safety problems, *International Civil Aviation Organization (ICAO) Journal*, Vol. 51, 1996, pp. 11–13.

Oliverira, A.M., Melo, A.C.G., Pinto, L.M.V.G., Consideration of equipment failure parameter uncertainties in bus composite reliability indices, *Proceedings of the IEEE Power Engineering Society Winter Meeting*, 1999, pp. 448–453.

Olivier, D.P., Engineering process improvement through error analysis, *Medical Device & Diagnostic Industry Magazine*, Vol. 21, No. 3, 1999, pp. 130–136.

Panzera, V.M., Managing operational safety in all phases of the life cycle of railway operations, *Proceedings of the 26th Annual Southern African Transport Conference*, 2007, pp. 801–811.

Parasuraman, R., Hancock, P.A., Olofinboba, O., Alarm effectiveness in driver-centred collision-warning systems, *Ergonomics*, Vol. 40, No. 3, 1997, pp. 390–399.

Parker, L.E., Draper, J.V., Robotics applications in maintenance and repair, in *Handbook of Industrial Robotics*, edited by S.Y. Nof, John Wiley & Sons, New York, 1999, pp. 1023–1036.

Pekka, P., Kari, L., Lasse, R., Study on human errors related to NPP maintenance activities, *Proceedings of the IEEE Conference on Human Factors and Power Plants*, 1997, pp. 12/23–12/28.

Pekkarinen, A., Vayrynen, S., Tornberg, V., Maintenance work during shut-downs in process industry-ergonomic aspects, *Proceedings of the International Ergonomics Association World Conference*, 1993, pp. 689–691.

Pesme, H., Le Bot, P., Meyer, P., Little stories to explain human reliability assessment: A practical approach of MERMOS method, *Proceedings of the IEEE Conference on Human Factors and Power Plants*, 2007, pp. 284–287.

Piggin, R., A new approach to robotic safety: Safety BUS p at BMW, *Industrial Robot*, Vol. 29, No. 6, 2002, pp. 524–529.

Ping, L., Xiufang, C., Reliability analysis of the stability of continuous welded rail track, *Journal of Structural Engineering (Madras)*, Vol. 27, No. 1, 2000, pp. 49–53.

Pontt, J., Rodriguez, J., Dixon, J., Safety, reliability, and economics in mining systems, *Proceedings of the 41st IEEE Industry Applications Conference*, 2006, pp. 1–5.

Pyy, P., Laakso, K., Reiman, L., A study on human errors related to npp maintenance activities, *Proceedings of the IEEE Conference on Human Factors and Power Plants*, 1997, pp. 12/23–28.

Rahimi, M., System safety for robots: An energy barrier analysis, *Journal of Occupational Accidents*, Vol. 8, 1984, pp. 127–138.

Raman, J.R., Gargett, A., Warner, D.C., Application of HAZOP techniques for maintenance safety on offshore installations, *Proceedings of the First International Conference on Health, Safety Environment in Oil and Gas Exploration and Production*, 1991, pp. 649–656.

Ramdass, R., Maintenance error management, *Proceedings of the European Aviation Safety Seminar*, 2006, pp. 2–4.

Randolph, R.F., Boldt, C.M.K., Safety analysis of surface haulage accidents, *Proceedings of the 27th Annual Institute on Mining Health, Safety, and Research*, 1996, pp. 29–38.

Rankin, W., Hibit, R., Allen, J., Sargent, R., Development and evaluation of the maintenance error decision aid (MEDA) process, *International Journal of Industrial Ergonomics*, Vol. 26, 2000, pp. 261–276.

Rao, A., Tsai, T., Safety standards for high-speed rail transportation, *Transportation Research Record*, No. 1995, 2007, pp. 35–42.

Rasmussen, J., Human errors: A taxonomy for describing human malfunction in industrial installations, *Journal of Occupational Accidents*, Vol. 4, 1982, pp. 311–335.

Reason, J., Approaches to controlling maintenance error, *Proceedings of the 11th FAA/AAM Meeting on Human Factors in Aviation Maintenance and Inspection*, 1997, pp. 9–17.

Reason, J., *Cognitive Engineering in Aviation Domain*, Lawrence Erlbaum Associates, Mahwah, NJ, 2000.

Reason, J., Corporate culture and safety, *Proceedings of the Symposium on Corporate Culture and Transportation Safety*, April 1997, pp. 187–194.

Reason, J., Hobbs, A., *Managing Maintenance Error: A Practical Guide*. Ashgate Publishing Company, Aldershot, 2003.

Reason, J., Maintenance-related errors: The biggest threat to aviation safety after gravity? *Proceedings of the International Aviation Safety Conference*, 1997, pp. 465–470.

Reinach, S., Viale, A., Application of a human error framework to conduct train accident/incident investigations, *Accident Analysis and Prevention*, Vol. 38, 2006, pp. 396–406.

Resor, R.R., Patel, P.K., Allocating track maintenance costs on shared rail facilities, *Transportation Research Record*, No. 1785, 2002, pp. 25–32.

Reynolds, R.L., History of coal mine electrical fatalities since 1970, *IEEE Transactions on Industry Applications*, Vol. 21, No. 6,1985, pp. 1538–1544.

Rognin, L., Salembier, P., Zouinar, M., Cooperation, reliability of socio-technical systems and allocation of function, *International Journal of Human Computer Studies*, Vol. 52, No. 2, 2000, pp. 357–379.

Rudd, D., Our chance to put rail safety first, *Professional Engineering*, Vol. 14, No. 19, 2001, pp. 17–19.

Ruff, T.M., Holden, T.P., Preventing collisions involving surface mining equipment: A GPS-based approach, *Journal of Safety Research*, Vol. 34, No. 2, 2003, pp. 175–181.

Russell, J.W., Robot safety considerations: A checklist, *Professional Safety*, December 1983, pp. 36–37.

Russell, S.G., The factors influencing human errors in military aircraft maintenance, *Proceedings of the International Conference on Human Interfaces in Control Room*, 1999, pp. 263–269.

Saccomanno, F.F., Craig, L., Shortreed, J.H., Truck safety issues and recommendations of the conference on truck safety: Perceptions and reality, *Canadian Journal of Civil Engineering*, Vol. 24, No. 2, 1997, pp. 357–369.

Sakai, H., Amasaka, K., The robot reliability design and improvement method and the advanced Toyota production system, *Industrial Robot*, Vol. 34, No. 4, 2007, pp. 310–316.

Sammarco, J.J., Programmable electronic and hardwired emergency shutdown systems: A quantified safety analysis, *Proceedings of the 40th IEEE Industry Applications Society Annual Meeting*, 2005, pp. 210–217.

Sanso, B., Mellah, H., On reliability, performance and Internet power consumption, *Proceedings of the 7th International Workshop on Design of Reliable Communication Networks*, 2009, pp. 259–264.

Schneidewind, N.F., Reliability modelling for safety-critical software, *IEEE Transactions on Reliability*, Vol. 46, No. 1, 1997, pp. 88–98.

Segerman, A.M., Covariance as a metric for catalog maintenance error, *Proceedings of the AAS/AIAA Space Flight Meeting*, 2006, pp. 2109–2127.

Sherif, Y.S., Kheir, N.A., Reliability and failure analysis of computer systems, *Computers and Electrical Engineering*, Vol. 11, 1984, pp. 151–157.

Shorrock, S.T., Kirwan, B., Development and application of a human error identification tool for air traffic control, *Applied Ergonomics*, Vol. 33, No. 4, 2002, pp. 319–336.

Sraeter, O., Kirwan, B., Differences between human reliability approaches in nuclear and aviation safety, *Proceedings of the IEEE 7th Human Factors Meeting*, 2002, pp. 3.34–3.39.

Stewart, B., Iverson, S., Beus, M., Safety considerations for transport of ore and waste in underground ore passes, *Mining Engineering*, Vol. 51, No. 3, 1999, pp. 53–60.

Stormont, D., Offshore rig designed for safety, *Oil and Gas Journal*, Vol. 57, No. 9, 1959, pp. 147–148, 151.

Sung, C., Development of optimal production, inspection, and maintenance strategies with positive inspection time and preventive maintenance error, *Journal of Statistics and Management Systems*, Vol. 8, No. 3, 2005, pp. 545–558.

Sykes, E.H., Mine safety appliances, *Canadian Mining Journal*, Vol. 48, No. 43, 1927, pp. 856–861.

Taylor, J.C., Reliability and validity of the maintenance resources management/technical operations questionnaire, *International Journal of Industrial Ergonomics*, Vol. 26, 2000, pp. 217–230.

Taylor, M., Integration of life safety systems in a high-risk underground environment, *Engineering Technology*, Vol. 8, No. 7, 2005, pp. 42–47.

Telle, B., Vanderhaegen, F., Moray, N., Railway system design in the light of human and machine unreliability, *Proceedings of the IEEE International Conference on Systems, Man and Cybernetics*, Vol. 4, 1996, pp. 2780–2785.

Thompson, P.W., Safer design of anaesthesia equipment, *British Journal of Anaesthesia*, Vol. 59, 1987, pp. 913–921.

Tien, J.C., Health and safety challenges for China's mining industry, *Mining Engineering*, Vol. 57, No. 4, 2005, pp. 15–23.

Toriizuka, T., Application of performance shaping factor (PSF) for work improvement in industrial plant maintenance tasks, *International Journal of Industrial Ergonomics*, Vol. 28, No. 3–4, 2001, pp. 225–236.

Ujimoto, K.V., Integrating human factors into the safety chain—A report on international air transport association's (IATA) human factors "98", *Canadian Aeronautics and Space Journal*, Vol. 44, No. 3, 1998, pp. 194–197.

Ulrich, K.T., Tuttle, T.T., Donoghue, J.P., Townsend, W.T., *Intrinsically Safer Robots*, NASA Report No. NAS 10-12178, Barrett Technology, Inc., Cambridge, Massachusetts, May 1995.

Underwood, R.I., Occupational health and safety: Engineering the work environment-safety systems of maintenance, *Proceedings of the International Mechanical Engineering Congress on Occupational Health and Safety*, 1991, pp. 5–9.

Urban, K., Safety in the design of offshore platforms: Integrated safety versus safety as an add-on characteristic, *Safety Science*, Vol. 45, No. 1–2, 2007, pp. 107–127.

Vanderhaegen, F., APRECIH: A human unreliability analysis method—Application to railway system, *Control Engineering Practice*, Vol. 7, No. 11, 1999, pp. 1395–1403.

Vanderhaegen, F., Non-probabilistic prospective and retrospective human reliability analysis method—Application to railway system, *Reliability Engineering and System Safety*, Vol. 71, 2001, pp. 1–13.

Varma, V., Maintenance training reduces human errors, *Power Engineering*, Vol. 100, No. 8, 1996, pp. 44, 46–47.

Vaurio, J.K., Optimization of test and maintenance intervals based on risk and cost, *Reliability Engineering and System Safety*, Vol. 49, No. 1, 1995, pp. 23–36.

Vlenner, C.A., Drury, C.G., Analyzing human error in aircraft ground damage incidents, *International Journal of Industrial Ergonomics*, Vol. 26, No. 2, 2000, pp. 177–199.

Wang, C.H., Sheu, S.H., Determining the optimal production-maintenance policy with inspection errors: Using a Markov chain, *Computers and Operations Research*, Vol. 30, No. 1, 2003, pp. 1–17.

Ward, M., McDonald, N., An European approach to the integrated management of human factors in aircraft maintenance introducing the IMMS, *Proceedings of the Conference on Engineering Psychology and Cognitive Ergonomics*, 2007, pp. 852–859.

Wattanapongskorn, N., Coit, D.W., Fault-tolerant embedded system design and optimization considering reliability estimation uncertainty, *Reliability Engineering and System Safety*, Vol. 92, 2007, pp. 395–407.

Weide, P., Improving medical device safety with automated software testing, *Medical Device & Diagnostic Industry Magazine*, Vol. 16, No. 8, 1994, pp. 66–79.

Weinger, M.B., Anesthesia equipment and human error, *Journal of Clinical Monitoring & Computing*, Vol. 15, No. 5, 1999, pp. 319–323.

Welch, D.L., Human error and human factors engineering in health care, *Biomedical Instrumentation & Technology*, Vol. 31, No. 6, 1997, pp. 627–631.

Wells, D.J., Hazardous area robotics for nuclear systems maintenance: A challenge in reliability, *Proceedings of the Portland International Conference on Management of Engineering and Technology*, 1991, pp. 371–373.

Wenner, C.A., Drury, C.G., Analyzing human error in aircraft ground damage incidents, *International Journal of Industrial Ergonomics*, Vol. 26, 2000, pp. 177–199.

Wenner, C.L., Drury, C.G., A unified incident reporting system for maintenance facilities, In *Human Factors in Aviation Maintenance-Phase VI: Progress Report*, Vol. II, Office of Aviation Medicine, Federal Aviation Administration, Washington, DC, 1996, pp. 191–242.

Wen-Ying, W., Der-Juinn, H., Yan-Chun, C., Optimal production and inspection strategy while considering preventive maintenance errors and minimal repair, *Journal of Information and Optimization Sciences*, Vol. 27, No. 3, 2006, pp. 679–694.

White, P., Dennis, N., Tylor, N., Analysis of recent trends in bus and coach safety in Britain, *Safety Science*, Vol. 19, No. 2–3, 1995, pp. 99–107.

Wiklund, M.E., Human error signals opportunity for design improvement, *Medical Device & Diagnostic Industry Magazine*, Vol. 14, No. 2, 1992, pp. 57–61.

Williams, J.C., Willey, J., Quantification of human error in maintenance for process plant probabilistic risk assessment, *Proceedings of the Institution of Chemical Engineers Symposium*, 1985, pp. 353–365.

Winfield, A.F.T., Nembrini, J., Safety in numbers: Fault-tolerance in robot swarms, *International Journal of Modelling, Identification and Control*, Vol. 1, No. 1, 2006, pp. 30–37.

Winterton, J., Human factors in maintenance work in the British coal industry, *Proceedings of the 11th Advances in Reliability Technology Symposium*, 1990, pp. 312–322.

Wood, B.J., Ermes, J.W., Applying hazard analysis to medical devices, Part II, *Medical Device & Diagnostic Industry Magazine*, Vol. 15, No. 3, 1993, pp. 58–64.

Woodward, J.B., Parsons, M.G., Troesch, A.W., Ship operational and safety aspects of ballast, water exchange at sea, *Marine Technology*, Vol. 31, No. 4, 1994, pp. 315–326.

Wu, B., Shi, H., The research of active fault-tolerant control for a ship propulsion system, *Journal of Harbin Engineering University*, Vol. 27, 2006, pp. 426–431.

Xie, M., Dai, Y.S., Poh, K.L., *Computing Systems Reliability: Models and Analysis*, Springer Inc., New York, 2004.

Zamojski, W., Caban, D., Impact of software and failures on the reliability of a man–computer system, *International Journal of Reliability, Quality and Safety Engineering*, Vol. 13, No. 2, 2006, pp. 149–156.

Zhongxiu, S., Reliability analysis and synthesis of robot manipulators, *Proceedings of the Annual Reliability and Maintainability Symposium*, 1994, pp. 201–205.

Ziskovsky, J.P., Working safely with industrial robots, *Plant Engineering*, May 1984, pp. 81–85.

Index